# MANAGER'S GUIDE
# TO MACHINERY MAINTENANCE
## A Master Plan
## for Organization and Control

# MANAGER'S GUIDE
# TO MACHINERY
# MAINTENANCE
## A Master Plan
## For Organization and Control

*Richard L. Weaver*

PRENTICE HALL, Englewood Cliffs, New Jersey 07632

Library of Congress Cataloging-in-Publication Data

Weaver, Richard L.,
    Manager's guide to machinery maintenance : a master plan for
  organization and control / Richard L. Weaver.
        p.   cm.
    Includes index.
    ISBN 0-13-553082-2
    1. Machinery--Maintenance and repair.    I. Title.
  TJ153.W38  1991
  621.8'16--dc20                                                   90-26507
                                                                      CIP

Editorial/production supervision
and interior design: *Carol L. Atkins*
Cover design: *Lundgren Graphics, Ltd.*
Manufacturing buyer: *Kelly Behr* and *Susan Brunke*
Acquisitions Editor: *Mike Hays*

 © 1991 by Prentice-Hall, Inc.
A Division of Simon & Schuster
Englewood Cliffs, New Jersey 07632

Printed in the United States of America
10  9  8  7  6  5  4  3  2  1

ISBN 0-13-553082-2

Prentice-Hall International (UK) Limited, *London*
Prentice-Hall of Australia Pty. Limited, *Sydney*
Prentice-Hall Canada Inc., *Toronto*
Prentice-Hall Hispanoamericana, S.A., *Mexico*
Prentice-Hall of India Private Limited, *New Delhi*
Prentice-Hall of Japan, Inc., *Tokyo*
Simon & Schuster Asia Pte, Ltd., *Singapore*
Editora Prentice-Hall do Brasil, Ltda., *Rio de Janeiro*

# Contents

# Preface

This book was written as a reaction to the obvious need for some sort of order to be applied to maintenance, especially as applied to machinery. It is hoped that the information and procedures set forth will enable those companies that want a workable system, or an improved existing one, to install a maintenance plan that will offer control over the service and repair activities. The author has focused on the problems involved with machinery and, in particular, its conservation.

Conservation may be a unique way to describe effective preventive maintenance but that is exactly what it is. Capital equipment forms the basic means by which a company produces its goods or services. If the equipment fails to function properly, the entire company suffers.

The machinery discussed in this book is taken to mean those units or components that can be defined as mechanical and that use fuels and/or lubricating oils. There are many industries that use this type of machinery and numerous types of equipment that fit this broad category. A sampling of industries would include trucking, railway, ship, and tug companies. Municipal bus and refuse removal organizations, the construction industry, mining, corporate agriculture, and port facilities are additional examples.

There has always been an atmosphere of mystery surrounding the inner workings of machinery. This is particularly true for managers who often look upon the maintenance organization as something needed but not always understood or trusted. The lack of understanding that leads to this element of mistrust is often caused by the fact that little is known about the mechanical aspects of machinery and even less about the necessary upkeep involved. Thus management often feels that it must leave the maintenance organization to its own devices and hope that things run smoothly. Those in management that do try to get a better understanding of their maintenance department often get lost when faced with the technical details of maintenance. One unfortunate result is that higher management often puts more reliance and faith in the mechanics and foremen than they do in the supervisory personnel within a maintenance organization. The maintenance manager often becomes the poor relation in the family of company managers. This is indeed unfortunate because machinery maintenance today must be considered a management function and not a repair function.

Experience accumulated over the years, dealing with organizations around the world, has provided the technical background for this book. It has also indicated a need for some sort of simple detailed plan for dealing with the varied requirements of machinery maintenance. This need for direction and control is not limited to developing countries. There are many businesses in the United States that need assistance in this field.

Throughout the book, we stress how to control the maintenance process with strict record keeping, controlled scheduled servicing, and continuous equipment monitoring. At the same time, we have tried to make the system simple enough so that it is understood by everyone involved. This includes higher managerial levels. In this way the mystery is removed from the maintenance process. The results gained from the system will provide savings throughout the maintenance organization and will also enable management to formulate plans based on accurate data and not guesswork.

In referring to the maintenance organization and personnel, we have used the terms *maintenance manager* and *maintenance management*. There may be cases where the maintenance department covers more than machinery upkeep, as, for example, a company that includes buildings and grounds in that organization. Large plant maintenance departments may thus include a subdivision in charge of machinery and mobile equipment. In this book, the supervisory levels, including that of maintenance manager, are taken to mean those people within the group who deal with machinery upkeep and repair.

Two people in particular have given the author technical help and encouragement with the writing of this book. The first is my wife, Alicia, who encouraged me to write a book based on my experiences in the field as a maintenance management consultant. She also provided much needed help putting together the manuscript.

Mr. Robert L. Kincaid has also been of tremendous assistance. He has been both a friend and a mentor in the field of mechanical systems integrity management. His firm, Spectron International, Inc., has provided many of the illustrations in Chapters 10 and 11. We have worked together on a number of projects and his experiences with various industries has provided valuable background information for the chapters dealing with mechanical systems integrity management and on-condition machinery monitoring.

Richard L. Weaver

# MANAGER'S GUIDE
# TO MACHINERY MAINTENANCE
## A Master Plan
## for Organization and Control

# chapter 1

# The Need for Maintenance Management

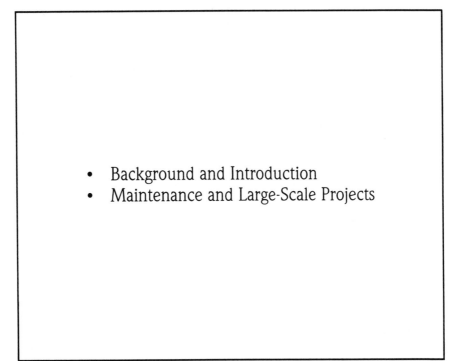

- Background and Introduction
- Maintenance and Large-Scale Projects

The need for strict management control has never been so great in the area of machinery maintenance. As with all other phases of business management, maintenance management has, at times, been sadly lacking in effectiveness. In certain types of industry, maintenance management barely exists, and it is virtually unknown in others. Maintenance generally runs the gamut between overkill and crisis. That is, equipment is periodically overhauled, whether needed or not, or repaired only as the result of failure. Neither of these extremes is satisfactory, as both are uneconomical and wasteful. In the first case, costly teardown and overhaul may not be necessary, and in the second instance, the cost of production losses must be added to the repair cost. Some firms go so far as to have a standby unit available in the event of a shutdown that might result in substantial production losses. Although there may be some instances where a backup unit is an absolute necessity, in general this only results in two units doing the job of one. This approach simply results in higher operating costs.

Since the end of World War II, the United States, and the free world in general, have been consumer oriented with little thought given to conservation of either natural resources or capital equipment. For example, during boom periods the construction industry would often include the full purchase price of bulldozers and backhoes in the bid price of a project. The contractor was not concerned about the condition of his equipment as long as it lasted for the duration of the job. During the period of high oil prices, the boom in the petroleum industry allowed that industry to operate with no thought given to the cost of equipment usage. The procedure was to let machinery run until it broke down, then repair or replace it without regard to cost. No wonder consumer prices were high. The public was paying for bad management.

When the economy is flourishing, these costs are simply passed on to the consumer as part of the price of producing the product. Government agencies collect more revenues and therefore pay for various public services, such as public transportation and waste removal. The general public never knows how much money is spent and lost as the result of poor maintenance and inefficient management. There is usually enough money available to provide adequate services to the public. As soon as the economy slows down, the quality of services declines. Municipal agencies clamor for new equipment so they can restore service to its proper level. The point to be made here is that with a good management team and a well-thought-out maintenance program, the equipment would give full service over a longer period regardless of the state of the economy.

How often have we heard or read in the news media that a city must replace its fleet of buses if proper service is to be provided? If a privately owned bus company were faced with the possibility of having to replace most of its fleet in a single year, it would be well on its way toward bankruptcy.

For the purpose of this book, we are considering industries that use heavy equipment. In general, we are talking about any machinery whose components require lubrication and periodic servicing. These include internal combustion engines, turbines, transmissions, reduction gears, compressors, and hydraulic components. Obviously, a good maninten ance program is going to take into account more than just these areas. However, these are the components that present the greatest potential for failure and the resulting high losses. At the same time, this is the general group that receives the poorest service. The list of industries that use this type of equipment include everything from airlines to mining, power-generating plants to the construction industry, mechanized agribusiness to port facilities, ships, railroads, light and heavy industry, the military, and other government agencies.

In a competitive market, good maintenance management can often mean the difference between profit and loss. Look at any successful large trucking company, and you will see that their equipment looks well maintained. Look beyond the glossy shine on the trucks and chances are you will find an efficiently organized maintenance department with a well-thought-out system of equipment maintenance and service. The individual systems may vary but all have one thing in common—a means whereby management knows the exact status of each unit and what its service requirements will be for both the short and long term. The alternative is usually chaos. It is a nightmare for those people whose job it is to keep the machinery operating if they do not have a good maintenance plan. More often than not, it boils down to crisis reaction and a mad scramble to put out fires.

The less-developed countries around the world are facing a real crisis caused in part by employing little or no maintenance in their operations. Over the years, international lending institutions have provided billions of dollars to third world countries to develop their resources, increase their agricultural output, provide public transportation and public works systems, expand port facilities, and develop fishing industries. These are but a few of the many projects being financed. In Africa, many countries have been independent for less than 20 years. After the newly independent nations had passed through the euphoric period of being free of their former colonial masters, they faced the task of building a workable economy. They had to develop their resources in order to export goods and obtain hard currency. No longer were they able to function under the former colonial system of trade.

In underwriting the various large-scale projects, the lending institutions counted on being repaid from the income derived from projected sales. On paper this looked good. One factor that was not given proper consideration was the lack of local expertise available to manage these large projects effectively. The advent of independence saw the exodus of many of the former colonists who had been the overseers and managers. There was a large-scale brain drain of the very people most needed to help get the new nations on their feet. The ensuing years have not allowed sufficient time to educate a new class of technicians and managers to assume all the duties required in the newly emerging economies. These economies are larger and much more complex than in colonial times.

Capital equipment has always been a major part of funding in any large project. Outside technical assistance was provided in starting up these projects and putting the basic elements in place, including facilities for servicing equipment. As the contract technicians turned the operations over to the local management staff, problems began to surface.

With the aid of hindsight, it is evident that some major omissions were made in preparing the local labor force to take over the maintenance of the capital equipment. In the first instance, the upper level of managers were not sufficiently trained and prepared for the job. Second, and more important, an overall system of maintenance management was not provided. The primary function of maintenance must be to control the care and use of equipment; it must not be thought of as just a repair function. Unfortunately, the latter is most often used as the primary goal.

The results of such shortsightedness are all too clearly visible. At some sugar estates, up to three-fourths of the agricultural and construction equipment was in a state of disrepair at the end of the second year of operation. Initially, mechanics and technicians were given instruction in the proper care of each piece of equipment by the manufacturer's representatives. Local maintenance managers sometimes, but not always, were included in this initial training. As personnel came and went, this specialized knowledge dissipated. Since there was no master plan to tie all the varied maintenance requirements together into one cohesive package, maintenance soon deteriorated into crisis repair management.

This is not an unusual situation. At some projects, equipment only three years old was consigned to a junk pile, later to be covered over with earth to provide landfill. This also helped hide the embarrassment of an obvious failure in the system. Undaunted, the governments of these countries would seek new loans to replace the lost equipment. They would put a new face on it by replacing one set of bureaucrats for another and claiming that the original equipment purchased was not adequate for the job. They would also point out that all the time, effort, and money already invested in the project would be lost without new replacement equipment. While all these arguments were true to a degree, the basic problem was not addressed. Amazingly, the loans were approved and the cycle repeated itself.

This problem is by no means confined to Africa. A study done at a major port facility in a South American country revealed that a few years after a major rehabilitation program was completed, the place was a catastrophe. Three-fourths of the capital equipment was inoperable. Not only that, it was not economical even to attempt repairs since the machinery had been so thoroughly cannibalized for parts.

At the port of New Orleans, an offshore work boat was tied up because of a major failure in one of its main engines. Investigation later revealed the cause to be a lack of lubricating oil in the crankcase. No one had bothered to make routine checks of the oil level in the engine. As a result, the boat was lost to service for weeks while expensive repairs were made to the damaged engine. It should be obvious that a simple plan for routine maintenance could have kept this boat in service and generating income.

Another offshore work boat undergoing a routine survey of its machinery was discovered to have dangerously corroded hull plates under the bilges. The owners were advised of this situation and were urged to have the vessel sent to a yard to have the paper-thin plates replaced. The overall condition of the vessel reflected that of the hull. Whether or not the owners intended to make repairs is not known. The vessel sank a week later enroute to an oil rig in the Gulf of Mexico. Survivors stated that even though it was night when they began taking on water in moderate seas, they were not concerned, as they were in sight of the rig and other vessels in the area. The pump could not handle the volume of seawater pouring in through the ruptured skin of the hull. Corrosion had also done extensive damage to the electrical system, so the radio was inoperative. The ship's whistle could not be used as a distress signal because it also was rusted.

These few examples serve to point out that there is something sadly lacking in industry. Maintenance is often considered a low-priority item in the overall scheme of things. It is the last to be considered in planning or budgeting, although actually, it should be given top priority. A piece of equipment is nothing more than a tool with which goods or services are produced. If it is not working at peak efficiency over a long, useful life, it can end up costing much more than it produces. Good maintenance will not necessarily ensure that a machine is used efficiently and to its full potential, but it will ensure that it will be kept mechanically ready and able to perform when required.

Lending institutions that finance large projects are beginning to realize that something has to be done to ensure that their investments are protected. Some are requiring that advance plans for equipment maintenance be submitted before requests for project loans are approved. Favorite subjects of these plans are outside technical assistance in caring for machinery, especially mobile equipment. Manufacturers are asked to provide factory technicians to set up the equipment and then stay on for six months to provide training and assistance in the operation and maintenance of the equipment. This adds to the initial cost and at best is short-term aid. It may look good on paper but does little more than delay the inevitable. A unified program must be established for the entire project, or fleet, that will continue in effect despite changes in personnel. Imported factory technicians will endeavor to maintain their company's equipment for the duration of the contract period and will train local people in the proper service and maintenance of the particular equipment. There is no attempt to train management in overall maintenance, service, spare parts storage, or cost control.

Local maintenance management staffs are left to fend for themselves in trying to bring things together into a workable overall plan. Starting from scratch can be an overwhelming task in the best of circumstances. Staffs are hampered by not getting expert guidance and are often led astray by directives and job descriptions issued by well-intentioned but ill-informed higher management. The result is almost certain failure.

This is not to imply that efforts have not been made to ease the situation or that the basic problem is being ignored. A lot of time and effort has been put into designing and building repair and service facilities plus training mechanics and technicians. Where we have not been successful is in the training and assistance that should be given to those at the lower and middle management levels. We have been discussing these problems primarily with respect to their effect on the less-developed countries, but many, if not most, are just as applicable here in the United States and in other highly industrialized nations.

Another factor that adds to the difficulties associated with operating a maintenance organization is the rapid advancement of technology and the complexity of the various makes of equipment produced. The technological changes incorporated in the newest equipment requires greater knowledge by repair and service personnel. Each unit has its own unique service and adjustments and these must be learned in order to keep the machinery running. Most of these technical advancements are incorporated to make the product operate more efficiently, particularly in the areas of operation and maintenance. Even in companies using more-or-less standardized equipment, maintenance requirements will vary.

Dozens of irrigation pumps driven by diesel engines may be used in a large corporate farming operation. Similarly, a tugboat company may utilize dozens of diesel engines for main propulsion and auxiliary services. Although each principal group of engines may be similar, differences in make, model, and year dictate differences in service procedures. Service intervals may differ, filters may not be the same, and the type of lubricating oil required may vary. It is easy to see why maintenance problems are so great and so often frustrating to those who have the responsibility of keeping the machinery running.

There is one other problem associated with machinery—failure through misuse. Failure of operators to use equipment with care and common sense, and misapplication of equipment by supervisors, often cause major damage, resulting in costly downtime and repairs. This is particularly disheartening to maintenance people who would otherwise keep the equipment in good working order. "They break 'em, we fix 'em," is a phrase often heard in a repair shop. Unfortunately, this is typical in what passes for a maintenance operation in many industries. Effective maintenance is much more than repairing equipment and putting it back into service. Repair is one thing, maintenance is quite another. The difference is that in the former, no control is being exercised over the care and use of machinery. Perhaps the most important task of a good maintenance program is to maintain control of all the equipment.

A good maintenance organization should have equal status with the other departments of a corporation, at least in relation to its size within the corporation. To function properly as a department and to make a lasting contribution to the corporation, it must have an established master plan for maintenance and service. This plan must function smoothly to bring all the various maintenance requirements of the equipment under one system that is operable and easy to understand.

Control and organization are perhaps the two most important elements in any successful plan. The remainder of the book is devoted to laying out, step by step, a system whereby a maintenance master plan is established and an organization is created to put it into operation. In addition, the subjects of preventive maintenance and on-condition monitoring are discussed in detail.

# chapter 2

# Organization

- Objectives
- Maintenance Organization

What constitutes an effective maintenance organization, and how is one established? These questions and others will be explained in detail throughout. As stated earlier, at the beginning the entire subject can seem to be overwhelming. It is too easy to get bogged down in details and lose sight of the main objective, which is to create a workable system that gives low- and mid-level management control over the operation and care of capital equipment. With this in mind, we concentrate on the supervisory and management levels rather than the hourly workers.

## OBJECTIVES

It matters little what descriptive title is attached to the group that is in charge of maintaining machinery: Maintenance Department, Equipment Division, Service Department, or even Plant Engineering. They all are involved in the same basic function, and that is to keep the equipment working in good order at the lowest possible cost. Cost represents the major factor that must be considered in operation of the maintenance department. By this, we do not mean simply the cost of repairs or the production losses incurred due to downtime. Repair cost is only one of many factors to be considered. Other expenses of running a department must be taken into account. For instance, there are fixed overhead costs, repair parts inventory, service items, and payroll expenses. Many, if not most of these costs can be controlled to a great extent. All these expenses become part of the overall cost of owning and operating the equipment. This, then, is the cost that must ultimately be controlled. Without this control, all will be lost. Thus the main objective of a maintenance organization must be to exercise strict control over the cost of the company's capital equipment.

Many organizations are content simply to collect data. Managers and supervisors sometimes spend a majority of their time compiling stacks of reports for submission to higher management. More often than not, these are after-the-fact statistics and do not reflect much more than the end result of an out-of-control situation. In reality, these statistics become more misleading as time goes on. Taken over a number of years, management will tend to believe that these data reflect a normal operating pattern for maintenance, especially if no noticeable fluctuations have occurred. If such data represent the only information available to higher management, on which to base its decisions, the company is in for trouble. The problem is that an unreliable data base is being generated. The data should accurately reflect the effectiveness of the maintenance organization and the efficiency of the equipment. Otherwise, management will be mislead into believing that a bad situation is normal and will base their planning accordingly.

The ultimate goal is control. To achieve that, there must be a plan for the maintenance organization that will include all the varied service and maintenance requirements brought together under a single system. It must also include the methods and personnel to administer the plan and maintain an accurate data bank.

## ORGANIZATION

Any maintenance organization should be responsible for more than just the repair and periodic servicing of machinery. This group should be in a position to provide technical assistance to management concerning matters dealing with capital equipment. It is only logical since this group should have the expertise and backup data. Thus it should be the most familiar with the mechanical details and economics of the equipment under its jurisdiction.

When it comes to replacing or adding capital equipment, this is the group to whom top management must turn to obtain the information essential for sound business decisions. If the maintenance group does not have accurate records and a firm control over its activities, the entire system breaks down.

To establish a workable organization, we must first define its major functions and then determine how best to create the required structure to provide the functions effectively. Personnel are responsible for:

1. Repairing equipment
2. Performing routine maintenance and servicing of machinery
3. Keeping detailed records of machinery and costs
4. Controlling costs of machinery upkeep
5. Training operating and maintenance personnel
6. Advising management on equipment and budgetary matters

The amount and cost of equipment repair are greatly influenced by the quality of routine service and maintenance given to that equipment. Machinery breaks down more from neglect than from accidents or misuse. Downtime caused by accidents and misuse can be reduced by training and educating operators and supervisors in the correct procedures to be applied to the equipment with which they will be involved.

The thing to remember is that a good organization is responsible for more than just patching up expensive machinery and shoving it out the shop door. To get the most out of both labor hours and machinery, complete and accurate records must be kept on each piece of equipment from the time it is purchased until it has reached the end of its useful life. In other words, we need to establish a complete machinery history for each unit.

If we can both develop a history and control the normal routine service and maintenance procedures, we will have a good start toward controlling costs. The objective is to have a simple workable system that monitors each unit's ownership and operating cost. Next, we want to be able to compare these costs to the income generated by each unit. This may take the form of hours worked, miles driven, units produced, or actual cash income received. It will then be possible to determine, among other things, when the unit begins to produce less than it costs to own and operate. The maintenance organization will then be in a position to take a look two or three years ahead and make realistic budgetary plans based on fact, not on wishful thinking.

All this will certainly provide greater cost control throughout the maintenance organization. It will also lower overhead costs by using labor hours and materials more efficiently. It is amazing what tight organization and controls will accomplish.

Finally, equipment can be analyzed on the basis of these data to determine if it performs as initially expected. It is possible that another size or model might do the job better and at lower cost. It may also be possible that a company has the correct equipment but too many units. Accurate analysis is impossible without good historical data.

What makes the difference between a good maintenance organization and a poor one should be evident. The ingredients of a good organization are:

1. Having a simple but workable organization
2. Managerial skills
3. Dedication
4. Planning

As with any other endeavor, maintenance should start with basics. In this case, the basic factors are the goals of a maintenance organization, which we have already described. The type of organization that is set up should be the simplest possible that can address these goals. It takes no special skill to set up a complicated system where every possible contingency must be covered by a person or group. This is known as a bureaucracy and is the first thing a maintenance organization should avoid.

There are two separate groups to be considered when setting up a maintenance organization. The first has to do with repair in shop, in the field, or at the work site. This group may be large or small, depending on the nature of the business or the type of equipment involved. The *repair group* will consist of supervisors, mechanics, electricians, welders, painters, and other skilled tradespeople in addition to other miscellaneous personnel. Since a large amount of the work of these personnel centers around repair shops, the day-to-day upkeep of shop buildings will also come within their jurisdiction. Clerks handling parts, payroll time, and so on, will also be included in this general group. There may be one person in charge of all the repair shops who reports to the department manager. Reporting to the repair shop manager will be the supervisors in charge of the various trade sections or repair areas.

The second group is the one concerned primarily with the routine maintenance and servicing of equipment. Since the work of this group is so important, the person in charge should be second to the maintenance department manager or, at the very least, be on the same level as the person in charge of the repair group. This is so because it is to this second group that most of the work of detailed record keeping and routine service is to be assigned. Because the group will be so intimately involved with records, cost data, and service scheduling, the person in charge should be a close advisor to the maintenance manager. These two people will be charged with making recommendations concerning equipment purchases or status changes based on the data available from the various records. Higher-level managers will make decisions based on this support information. This group, which we will call the *service group*, should also be assigned the task of training personnel in the service, operation, and correct application of equipment.

The service group includes the personnel who provide lubrication maintenance in the shop, in the field, or at the workstations. It will have all the equipment required to do the job, such as service stations, lubrication service trucks, fuel trucks, and portable equipment. The group will also include clerks and a supervisor whose function will be to collect and assemble the service and repair data on each unit, as well as operating hours, in order to schedule the required service on a timely basis. The supervisor has the added function of ensuring that all the various activities are coordinated and work together smoothly. The service group will issue service work orders for equipment based on the manufacturer's recommendations. To do this, the group must keep a running total of the actual operating time, mileage, or fuel consumed. Service is normally based on one of these rates.

The service group has the all-important task of overseeing the preventive maintenance of the machinery and in so doing must be fully aware of the current status of each unit under its care. This is the reason such strong emphasis is placed on accurate record keeping. The key word here is *preventive*. By preventing minor problems from developing into major breakdowns, losses due to downtime and repairs can be drastically reduced. This is the goal of good maintenance practice. It is amazing how often this simple truth is overlooked.

In the ideal situation, the service group's efforts should result in substantially reduced repair work and a corresponding reduction in the size of the repair group. There will always be a need for repairs, but the objective should be to reduce the volume to an absolute minimum. Except for breakdowns caused by accidents or other unforeseen events, the service group should be able to predict the need for repairs and schedule them in such a way that production losses and repair costs are kept to a minimum level. This approach to maintenance can practically eliminate catastrophic breakdowns resulting from neglect.

Figure 2.1 shows the basic outline of an effective maintenance organization. The chart shows only the essential elements. Different types and sizes of companies may require modifications to the basic structure. The point to remember is: *Keep it simple.* Do not allow the organization to mushroom into bureaucratic complexity, or that key element—control—will be lost.

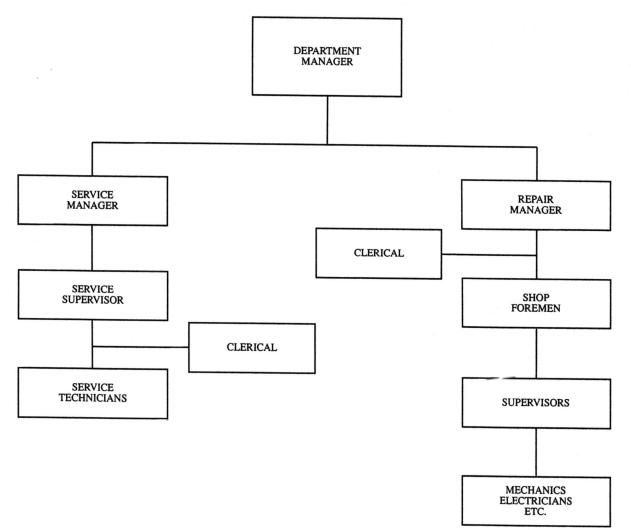

**Figure 2.1**  Basic organization chart for a maintenance department.

# chapter 3

# Personnel

- Repair Group
- Service Group
- General Administration

We have discussed the organizational requirements of good maintenance, but what about the people who are to fill the various key positions? Unfortunately, improper or poorly thought out staffing can cause the ruin of the best organizations. What might otherwise be a well-designed plan for administering a maintenance system can end up with disastrous results because the wrong people were assigned to staff positions.

This discussion will be, for the most part, concerned with filling supervisory positions with the most suitable people. For those in the various skilled trades, the required proficiency levels are easily determined. Requirements for the different levels of mechanics, electricians, welders, and so on, will vary with the type and size of company. As outlined in Chapter 2, there are two major groups in a maintenance organization. They are the repair group and the service group.

## REPAIR GROUP

The supervisors and foremen in the repair group should be people trained in one of the trades, having reached the highest level, such as master mechanic and master electrician, who have shown an aptitude for leadership. They should also have spent enough time with the company to become thoroughly familiar with the equipment they will be working with and repairing. Leadership is a vital part of any supervisory job, no matter what level is being discussed. It is the ability to inspire motivation in others and to direct this motivation toward accomplishing a task in an expeditious manner. As the word implies, *leadership* means to lead. It might be better thought of as meaning *to guide*. A foreman or supervisor does not necessarily have to be the best mechanic or electrician in the shop but should be the best at human relations. He or she must be able to see beyond a single task and coordinate the work within the assigned group. It should not have to be said, but the supervisor must have the ability to read and write. Too often, competent craftsmen find it difficult to cope with the written word.

The person in charge of the repair group may be described by any number of titles. "Repair manager," "shop supervisor," and "general foreman" are some of the terms used. Each type of company may have its preferred usage. For consistency in this book, *repair manager* will be used as the job title. Since the person in this position must oversee the various foremen and supervisors, he or she must possess the same qualities and abilities as they do, but with a difference in emphasis. For instance, the repair manager does not necessarily need to be as skilled in one of the trades but must be a more skillful leader and communicator. In addition, the manager has the added responsibility of administration. For this alone, he or she will need skills acquired both from experience and schooling.

There may be people who, over the years, have come up through the ranks and have the ability to function effectively as repair managers. Somewhere along the path of advancement, they need to be trained in the arts of organization and administration. This may be in the form of specialized night courses or in-house training given by the company. The latter should be an ongoing procedure and applies to all members of the organization. As a matter of good management technique, supervisors should always give guidance and direction to the people reporting to them.

The other source of repair manager personnel might be college-trained people who have engineering degrees. An engineer with a few years of practical experience is well equipped to handle this type of position, since he or she has the technical background to appreciate the problems involved and has the basic training to adapt easily to the administrative side. Again, the size and type of company will more or less dictate how this position is filled.

## SERVICE GROUP

At the lower end of the organization ladder are the service technicians. These are the people directly responsible for providing periodic maintenance and service to the equipment. From experience, ex-mechanics are best suited to this work. They do not have to be expert or masters at their craft. In fact, lower-level mechanics will do just as well if they are responsible and interested in keeping machinery running. Basically, those who appreciate the benefits of preventive maintenance and prefer working in different areas rather than being confined to a single workstation are the people to be recruited. They must also be able to work well under limited supervision. These personnel must be literate so they can follow written instructions and enter accurate data on service forms.

The clerks employed by this group must be first-class people who are accurate and conscientious. They must be willing and able to go beyond compiling data. They must be able to stir people into action when reports and data are not received on time. A key element of good preventive maintenance is timely and accurate data.

There must be a *service group supervisor,* who is placed in charge of the technicians and clerks. This person's duties will include issuing the special work assignments to the service personnel and ensuring that specialized periodic service and lubrication maintenance are done when required. The service group supervisor will ensure that status boards are kept up to date for all equipment. These boards are the means by which the various forms of routine maintenance are triggered at the appropriate times. (Status boards are discussed in detail later in the book.) The supervisor will see that the detailed information developed is collected and condensed onto the correct forms and is eventually entered on the maintenance history card of each unit.

The supervisor's other responsibilities include seeing that new equipment is properly introduced into the system and making adjustments to the system as equipment status changes. Finally, the service group supervisor assists the service manager in the day-to-day administration of the group. The person filling this position should at least have a high school education and have a mechanical aptitude so that the technical aspects of the work are familiar. Generally, one clerk or technician stands out in the organization, and this is the person who should get the job. Usually, it will be an older person who accepts responsibility and has previously performed well in a technically oriented job. The position might also be filled by a recent graduate of a technical school. Considerations for filling this position will be different from company to company, but the basic requirements remain valid.

Ideally, those selected for the positions of *service manager* and *maintenance department manager* should be engineers. This, of course, may not be practical for smaller organizations and may require modification for others. The junior position should be filled by someone who believes that good preventive maintenance is essential and that complete accurate records must be kept to ensure that management has the information it needs. The service manager must have analytical talents to interpret the data collected and take appropriate action to see that personnel and machinery operate efficiently. He or she must also act as an advisor to the maintenance department manager. Data collected by the service group should provide accurate information on cost and frequency of repairs, type of repairs, fuel and lube costs, operating times, availability, and chronic problem areas. These data should be available on a per unit basis as well as by group and for the complete inventory of equipment.

The service manager should assist the maintenance department manager with department and capital equipment budgets that are based on historical information. Accurate forecasts and comparisons of equipment can be made to help management make sound decisions concerning equipment and operations. Good planning by management is crucial to a healthy competitive company. If control is poor and the operating data are faulty, management cannot function properly and will usually make costly errors in planning that will have damaging results, perhaps for years.

The maintenance department manager needs someone who can be depended on in the position of service manager. The person must be technically oriented—must be able to grasp the technical details and service requirements of the equipment and the theory behind them. An active curiosity about machinery operation will make it natural to want to discover the cause and solution to problems. There may be occasional or chronic problems to deal with and these may involve a single unit or an entire group. A service manager must be able to isolate and remedy them. The person must be well organized, with a logical mind, and be skillful at planning. In overseeing the collection and compiling of information, he or she will be establishing a truly accurate data base—the foundation on which an effective maintenance system is built.

The service manager must be a good communicator because, among other things, the person must be able to train personnel in the proper service and use of the equipment. Technicians and clerks in the service group must be instructed in the proper procedures to be followed for each new unit and refreshed periodically on the older units. Operators of machinery also need instruction on new units and refresher training on older units. In addition to normal operating procedures, they must be educated in the operating characteristics, such as limits of temperature, pressure, and speed. They must know the importance of making visual inspections and reporting even slight discrepancies. The world's best service program cannot provide protection against faulty or negligent operation. An operator who ignores leaks, noises, or other indications of malfunction, and runs equipment until it fails, is the second major cause of breakdown after poor maintenance. It must be the responsibility of the service group to train such people. Then, if operators do not perform correctly, they should be replaced. It is a lot less expensive to replace an operator than to make major repairs or replace a major piece of equipment.

## GENERAL ADMINISTRATION

The maintenance department manager has overall responsibility for both repair and service. The manager must therefore be an administrator and is dependent on two group managers for assistance. He or she must have a technical background and also have well-developed management skills. As a member of the next-higher level of management, the maintenance manager's primary responsibility is to make sure that the organization is efficient and cost-effective. The department manager is the main link between maintenance and all the other company departments and thus must be an effective communicator. The manager's technical contribution to the company should be based on a solid background of experience and reliable data. The effectiveness of the manager's input will have a great influence on corporate planning, both short term and long range.

The terms *maintenance organization* and *maintenance department* have been used as general descriptive titles in referring to the group charged with the responsibility of keeping the corporation's major machinery operating at the lowest possible cost. It should be obvious by now that this involves much more than simply running equipment until it stops, then fixing it. Regardless of the terms used to describe the organization, the manager must be able to administer it effectively. To do this, he or she needs good people and a workable organization. A good manager should be able to set up a simple workable system, starting with the basic chart illustrated in Chapter 2. He or she must be able to attract and retain top-notch people to fill key positions. A manager who develops skills in human relations will find this part of the job much easier and more rewarding. With the necessary technical capability and leadership, based on example and guidance, a manager can put together a staff that will work together smoothly and effectively.

Organization charts may change due to varying circumstances and personnel will also change, for any number of reasons. What is needed is an organization master plan that will not change appreciably over the years. A master plan is an integral part of the organization. The term *organization* is normally thought of as describing staff positions and their relationship to each other. A master plan should fill in the gaps by adding the *how* to the *who*.

Having an understandable and workable master plan in place will allow the entire organization to operate smoothly. It tends to structure the work along definite avenues, all leading to a common goal. The goal here is twofold: to ensure that proper service and maintenance is performed on time and to collect all data pertaining to repairs, maintenance, service, and operation and condense it into a useful individual history of each piece of equipment.

The details of setting up a master plan are explained step by step later in the book. For now, it is enough to say that it is the heart of a good maintenance system. It consists of a simple straightforward system of reports, data sheets, files, instructions, and displays, all designed to manage preventive maintenance effectively and to give management continual statistical information concerning capital equipment. This is therefore the most important item for the maintenance manager to implement. The end result of an effective master plan is continuous up-to-date information on the status of each piece of equipment, groups of similar equipment, and the entire fleet or equipment inventory.

With a workable organization set up and a good master plan put into use, overall supervision and control become much easier at all levels. Everyone will understand how the system is supposed to work and so can oversee the various operations more effectively. Stress can be put on those areas that need the most attention and still allow enough time to stand back and observe how the overall system is functioning. Rather than being in a continuous state of harassment, trying to put out fires, managers can take the time to examine the problem areas and take corrective action. They will also build up a reliable data base for use as a reference when making decisions concerning equipment and procedures. Before describing a typical master plan, some further discussion is needed concerning the service group and its function within the maintenance organization.

# chapter 4

# The Service Group and the Master Plan

- Equipment Inventory List
- Routine Lube Service and Maintenance
- Equipment File Folder
- Maintenance History Card
- Fuels and Lubricating Oils
- Training
- Manuals and Publications
- Repair Data
- Lube Oil Monitoring

The service group's main function is to provide preventive maintenance, whereas the repair group is concerned primarily with restoring damaged, or failed, equipment to working order. As the preventive maintenance program increases in effectiveness, the amount of downtime for major repairs will decrease dramatically. A corresponding reduction in the size of the repair group can be expected.

A master plan must provide the framework on which a maintenance organization bases its operating procedures. Since any plan can include only those work functions that can be controlled, we come back to the term *preventive maintenance*. Thus the service group should have the responsibility for the administration and implementation of the master plan.

At this point it is a good idea to bring up the subject of computers, if only to make the point that they are another timesaving tool. As the method of setting up a master plan is explained step by step, it will become obvious that a computer would be a great aid in the collection and tabulation of data. From the standpoint of a maintenance organization, a computer can be a valuable aid in making comparative studies, in addition to being a data bank. The size and type of company will determine what, if any, computer assistance is available. Some may have a central computer with terminals in each department. In other cases, a small desktop personal computer will do the job. Still other companies may not have a computer of any kind, and for this reason, the establishment of a master plan will be explained using regular forms and files. Another important point to consider is that it is better to master a system manually before going to computer usage. It will be difficult enough to establish a master plan without adding the problems of computer programming and operating at the same time. Once the plan is in place and working, personnel will quickly get used to the routine. Then, if a computer is available, it will be much easier to program. Personnel will know exactly what they want from a computer and have the experience to have good programs made that will not only save time but will produce good results.

The service group is so important to an effective maintenance organization that its major areas of responsibility should be listed and explained in more detail. Some areas described here may not apply to all types of companies, but listing them will give the reader a better idea of the potential scope of control that can be realized by a good service group.

## EQUIPMENT INVENTORY LIST

A complete current listing of all equipment that falls under the jurisdiction of the maintenance organization must be made by the service group. This is the first step in creating a master plan. To function with any degree of success, this information must be compiled. An equipment list should contain the following information on each unit:

1. Assigned company number
2. Make, or name, description
3. Model number
4. Serial number
5. Model year
6. Year acquired
7. Rental rates (if applicable)

Machinery should be grouped together by make or type for easy classification. For example, all D-8 Caterpillar tractors might be in a single group if there were enough of them in a construction or agricultural fleet. If not, the group might have the heading "Caterpillar Crawler Tractors," where there are only a few of various models. Other examples of type grouping might include: Tractor-Trailer Trucks, Buses, Main Engines, Engine-Driven Generator Sets, Irrigation Pumps, Cranes, Ore Haulers, Gas Turbines, and Reduction Gears. Whatever the type company, similar units of equipment should be grouped together for quick identification and classification.

Each unit should be assigned a company number. Similar units should be assigned consecutive numbers or sets of numbers. Kenworth trucks might use numbers from 500 to 550. GMC trucks might be assigned numbers 551 through 599. By numbering in this manner, it will be easier to identify and describe a single unit. As soon as everyone becomes accustomed to the system, it will be obvious that it is easier to describe a unit by its number than by calling it the 1983 yellow Kenworth cabover with serial number such and such.

It is recommended that the company use a three- or four-digit numbering system with the numbers running from 100 to 999 or from 1000 to 9999. The amount of equipment owned and accounting practices will determine the best setup to be used. The company number also becomes an account number against which charges for repairs, fuel, lubricants, time, and miscellaneous consumables can be charged. Thus each major piece of machinery has its own account number and therefore can be monitored individually. As has been stated before, the object is to be able to monitor each piece of equipment individually and thereby have an accurate data base from which to evaluate one unit or the entire system. This phase of the implementation will require a close working relationship with the accounting department to set up the new account numbers. If not already in use, general account numbers must also be assigned. These would cover those departmental costs that cannot be directly charged to a unit of major equipment. Examples include repairs to shops, battery repair, utilities, salary payroll, and office supplies.

Financial and accounting people sometimes like to work with cost centers for fiscal accountability. Although this may be a useful approach for most departments, it does not necessarily work where a fleet of equipment is involved. To simply show that a particular tractor group cost a certain amount to repair and operate for a given period does not mean much. It is a lump-sum figure that cannot be compared with anything except the last report on that group. It is not an accurate indicator of the effectiveness of that machine grouping or of the department as a whole. However, if cost data are available on every individual unit, there is a good basis for comparison. The objective is to list all costs and expenses separately for each unit. These data, together with the time worked, or miles run, can be compared with data from other units. The totals for the group would equal those of a cost center. However, now the statistics are available to determine why costs are what they are for that period. Using the cost center method, it may be unclear why costs are high. By checking the individual costs, it might show that one or two units have had chronic problems and high instances of repair and so have caused the total group cost to increase. If the trend should continue, these units may be candidates for replacement. It might also show that the units operated significantly more than in the preceding reporting period and the increased costs may only reflect a normal increase in fuel and lube oil consumption. The unit numbering system is vital for cost accountability. Depending on the basic structure of the company, the unit numbers might be used as subaccount numbers to be added to the general account number of the maintenance department.

The first item on the equipment list is the unit's company number, followed by the make or name description (Fig. 4.1). This might be Caterpillar, GM, Ford, Dorman, and so on. A short one-word description of the make is all that is needed, and sometimes initials are sufficient; everyone recognizes GE, GM, B&W, and JD.

The model and serial number should be those of the complete unit and not a major component. Model year and year acquired are self-explanatory. Some companies rent out their equipment during slack periods, and others establish internal rental rates as a means of determining the amount of work the unit produces. The equipment inventory can double as a rental schedule.

The equipment inventory list maintained by the service group is a vital part of any master plan, as it provides the basis for the rest of the plan. The service group should be responsible for assigning the company numbers. Sufficient unassigned numbers should be left in each grouping to allow for the addition of new machinery. Making an equipment inventory list is the first step, and perhaps the most important one, in setting up a preventive maintenance master plan.

## *ROUTINE LUBE SERVICE AND MAINTENANCE*

Routine lube service and maintenance will take up the most time and effort of the personnel in the service group. In fact, almost all the activities of the service group are closely tied to routine lube service and maintenance. Another way of describing this is preventive maintenance.

To function properly, the service group must be thoroughly familiar with the service and operational requirements of each unit under its jurisdiction. As soon as a new unit is acquired, service personnel must record all pertinent data, such as the serial number of the unit and the make, model,

```
                        EQUIPMENT INVENTORY LIST

    CO. NO.    MAKE      MODEL     SERIAL NO.      MOD. YR.    YR. AQUI'D

                            IRRIGATION PUMPS

     100       Perkins    6-354     88756439       1975        1975
     101       Cat.       3160      6572331        1975        1975
     102       Cat.       3160      6572389        1975        1975
     103       Waukesha   6 WAKB    23866392       1975        1976




                         CATERPILLAR EQUIPMENT

     150       Cat.       D-5       94J657         1975        1975
     151       Cat.       D-5       94J924         1975        1975




                            PICKUP TRUCKS

     200       Ford       F250      455673BPQ125   1983        1983
     201       Ford       F250      457628BPR788   1983        1983
     202       Ford       F350      663476FTW435   1983        1983
     203       Ford       Bronco    7886CB23935    1983        1983




                            HEAVY TRUCKS

     250       Kenworth   552       61820          1953        1956
     251       Kenworth   552       61881          1953        1956
     252       Kenworth   552       61987          1953        1957
     253       Kenworth   552       68356          1957        1960
     254       Ford     LN 9000     R902VT37121    1974        1974




                            WHEEL TRACTORS

     300       County     1254      29216          1975        1975
     301       County     1164      32634          1976        1976
     302       County     1164      32537          1976        1976




                              1 of 5
```

**Figure 4.1**  Equipment inventory list.

and serial numbers of its major components. A company number is assigned and the unit is added to the equipment inventory list.

## *EQUIPMENT FILE FOLDER*

As soon as a unit is added to the equipment inventory list, a file is started on the unit. An equipment file folder is a legal-sized manila file folder with space provided on the outside for listing all pertinent information concerning that unit (Fig. 4.2). It is important that service personnel collect all the information available as soon as a unit is received. It is much easier to record data on a new, clean machine than to crawl over equipment that is already in use, hoping nameplates are in place and legible. It must be a matter of policy that no new equipment is put into service until it has been logged in, inspected, fueled, lubed, and assigned a company number. In addition to costs already discussed, operator's time is often charged to the unit, so it must have a number before it is put to work.

Once the details of the equipment data have been transferred to a file folder, it becomes a permanent record. These data are of great value when component repairs are needed. The model and serial numbers will be needed in order to determine warranty status, obtain factory technical assistance, or get the correct repair parts. The amount of information required for the folder may vary according to the type of machinery used by the company. Figure 4.2 illustrates a folder that might be used for construction or agricultural equipment. It might also be used for fleets of buses, trucks, or other vehicles. In addition to the various nameplate data, the service and operator's manuals should be studied to determine the type of lubricants required, filter elements used, tire sizes, belt types, and any other information that might be needed on short notice. There are enough blank spaces available on the folder for any information considered useful. These individual file folders are usually filed, in order, by company number. Often, the file is set up by group and number so that it is in the same sequence as the equipment inventory list.

The equipment file folder and its contents constitute a permanent record. The record remains active until the unit is disposed of, in which case it can be destroyed or, if the unit is sold, can go with the unit as a valuable record for the new owner.

## *MAINTENANCE HISTORY CARD*

The next item to attend to is the maintenance history card (Fig. 4.3). One card is made for each unit. Initially, only the top front of the card is filled out. Each card records the history of its unit for one year, so the year is entered together with the unit's company number, make, model, serial number, and other details. The front of the card has space for monthly totals of the various costs of fuel, lubricants, repair parts, and repair labor costs. These are entered as dollar figures and the totals for the year are tabulated at the bottom. It is a good idea to enter the equipment cost and depreciation

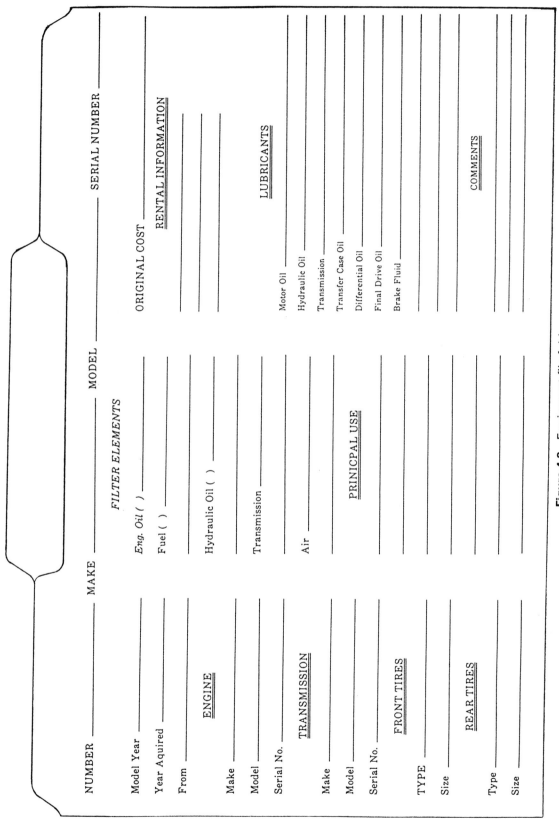

**Figure 4.2** Equipment file folder.

**Figure 4.3** Maintenance history card: front.

EXPLANATION OF REPAIRS

YEAR _____

| MO. | HOURS/MILES | ENGINE | TRANSMISSION | FINAL DRIVES | ELECTRICAL SYSTEM | HYDRAULIC SYSTEM | OTHER |
|---|---|---|---|---|---|---|---|
| 1 | | | | | | | |
| 2 | | | | | | | |
| 3 | | | | | | | |
| 4 | | | | | | | |
| 5 | | | | | | | |
| 6 | | | | | | | |
| 7 | | | | | | | |
| 8 | | | | | | | |
| 9 | | | | | | | |
| 10 | | | | | | | |
| 11 | | | | | | | |
| 12 | | | | | | | |

**Figure 4.3**  Continued (back of form).

information, as it will be valuable data for future reference. The reverse side of the card (Fig. 4.4) is self-explanatory. It helps to have a very brief description of repairs made over the years, as it is almost impossible otherwise to have accurate historical repair data readily available for study. This makes it easy to observe trends, in either a single unit or a group of similar units. After a year or two, this information will be invaluable for future planning.

When initially establishing a master plan, it will be necessary to visually inspect every piece of equipment to be included in the plan. All nameplate data that are available, plus other pertinent information, must be recorded and transferred to a master equipment inventory list, equipment file folders, and maintenance history cards. This will take time, and collecting as much data as possible on existing equipment will require the cooperation of other departments.

This is the heart of a good preventive maintenance program. The service group will expand this basic setup to include the procedures and format for carrying out the routine steps in the preventive maintenance program, as explained in detail in succeeding chapters.

## FUELS AND LUBRICATING OILS

In many organizations, the service group is given control of all functions having to do with fuel and lubricants, sometimes including purchasing. It is extremely important that rigid control be exercised over these products. One obvious reason is the cost factor. An accurate accounting of exactly how much fuel and lube oil is dispatched to each piece of equipment is necessary to assign accurate cost figures to those units.

The service unit should be in charge of keeping inventories of the bulk storage areas for fuel and lubricants. In addition, service stations, lubrication service trucks, fuel trucks, and portable equipment must come under the direct control of the service group. In the administration of fuels and lubricants, the service group must keep the maintenance manager and purchasing department constantly informed as to inventory levels through the use of monthly reports.

By conducting monthly inventories, service personnel will know the total usage for the period and exactly how much is on hand. One of the common breakdowns in internal communications results in delays in fuel and oil deliveries. Having equipment idle for lack of fuel or proper lubricating oil can be avoided by establishing simple controls. Knowing the total usage of fuel allows a comparison to be made with the cumulative totals dispatched to the equipment and for miscellaneous usage during the month. There will be some minor differences, but discrepancies due to pilferage or poor record keeping can be identified and corrected.

The service unit should also be mindful of the quality and cost of the fuel and lubricants used throughout the operation. It pays to look for the best price, but quality must not be sacrificed in the process. Service manuals will give general guidelines to follow when choosing lubricants. Some manufacturers will go so far as to name a particular brand and type. All major oil companies offer products that meet equipment manufacturers' specifica-

EXPLANATION OF REPAIRS

YEAR _____

| MO. | HOURS / MILES | ENGINE | TRANSMISSION | FINAL DRIVES | ELECTRICAL SYSTEM | HYDRAULIC SYSTEM | OTHER |
|-----|---------------|--------|--------------|--------------|-------------------|------------------|-------|
| 1 | | | | | | | |
| 2 | | | | | | | |
| 3 | | | | | | | |
| 4 | | | | | | | |
| 5 | | | | | | | |
| 6 | | | | | | | |
| 7 | | | | | | | |
| 8 | | | | | | | |
| 9 | | | | | | | |
| 10 | | | | | | | |
| 11 | | | | | | | |
| 12 | | | | | | | |

**Figure 4.4** Maintenance history card: back.

# MAINTENANCE HISTORY CARD

YEAR _____

Make _____  
Model _____  
Serial Number _____  
Model Year _____  
Co. Number _____  
Date of Acquisition _____  

Original Cost _____  
Depreciation _____  
Current Book Value _____  
Principal Use _____  

| MO | COST OF FUEL AND LUBRICANTS | | | | | | | REPAIR COSTS | | | |
|---|---|---|---|---|---|---|---|---|---|---|---|
| | DIESEL | GASOLINE | ENGINE OIL | TRANS. OIL | FINAL DRIVE OIL | HYD. OIL | TOTAL COST FUEL & LUB. | LABOR | MATERIALS | TOTAL COST REPAIRS | TOTAL |
| 1 | | | | | | | | | | | |
| 2 | | | | | | | | | | | |
| 3 | | | | | | | | | | | |
| 4 | | | | | | | | | | | |
| 5 | | | | | | | | | | | |
| 6 | | | | | | | | | | | |
| 7 | | | | | | | | | | | |
| 8 | | | | | | | | | | | |
| 9 | | | | | | | | | | | |
| 10 | | | | | | | | | | | |
| 11 | | | | | | | | | | | |
| 12 | | | | | | | | | | | |
| Total | | | | | | | | | | | |

**Figure 4.4** Continued (front of form).

tions. Where practical, it is a good idea to standardize as much as possible. Limiting the number of different lubricants will reduce the risk of using the wrong product by mistake and will increase the company's buying power. It is usually less expensive to buy a few products in large quantities than a number of varied ones in small lots. There are a number of publications available that are useful in helping to choose lubricants. Dealers for major oil companies can be very helpful in advising which of their products meets, or exceeds, the manufacturer's requirements.

## TRAINING

The service group must ensure that its personnel are kept up to date on the lubrication service and routine maintenance requirements of each unit. Service must also be familiar with each unit's operational requirements and limitations. To accomplish this, the service manager and supervisor should schedule training sessions throughout the year. These sessions should include the people who operate the equipment and personnel from the repair group as well as service employees. All should be familiar with each piece of equipment. The operating personnel are included so that they can learn about the special features of the machinery they normally use. They must be made to feel that they are an important part of a team and share responsibility for keeping the equipment running properly. Repair personnel are included for the same reason. They will also be in charge of performing routine service adjustments and mechanical checks on various system components.

The major topics covered in these training sessions are described below.

### Operator Training

Each person should be aware of how much fuel, water, lube oil, hydraulic oil, and so on, a unit is designed to hold. Just as important is the method of checking levels and how and when to make these checks. Operators should routinely make daily checks of their equipment. Normal operating data must be understood. Items such as normal coolant temperature, oil pressure, operating speeds, and air pressure are examples of operating data that must be understood. Operators must be thoroughly trained in the proper operating procedures so they will know when a machine is not performing properly and take the appropriate action.

Much of this training should be with the other two groups present. The purpose is to instruct operators as to why and when good lube service and routine maintenance must be given to their equipment. It also serves to assure them that they are valuable members of a team with a common objective. When provided with the proper basic training, operators will understand their equipment much better, give better care, and produce more at less cost.

The actual operation of the units should include emergency shutdown procedures. These would be triggered if any of the operating limits were to be exceeded. Operators must be trained to keep an eye on all gauges and understand what they mean. They should be instructed on how to make checks of fluid levels on a daily, or shift, basis. Visual inspections should be made to detect leaks, loose fittings or parts, and anything else that might indicate abnormal situations. This includes strange sounds and erratic movements. If something is not right, service personnel should be notified and the unit shut down. It is important to understand the emergency shutdown procedure for each piece of equipment in order to prevent damage to it or associated machinery. Above all, operators must be held responsible for their machines. For a system to be fair, however, operators must be given the proper instruction. Their supervisors should be included in these sessions so they will be aware of the procedures required.

### Service Group Training

The service group must be made aware of the schedules and locations for lubrication service and routine mechanical adjustments. There are grease points to learn as well as the location of fill points for various fluids. Types and quantities of fluids must also be learned. As will be shown later, it is not necessary to memorize all this information, as some is included in periodic service sheets and standard instructions. These people have to be motivated to appreciate the need for proper and timely routine maintenance as a method of keeping machinery running smoothly.

### Repair Group Training

Repair group training must be coordinated with the repair manager. With new equipment especially, the manufacturer often offers repair and maintenance instruction. It might be given at the dealer's place of business or at the owner's, or both. The service unit should schedule these sessions so that both service and repair groups can take full advantage of them. The training usually includes lube service and routine maintenance. There are usually numerous periodic adjustments to be made to the machinery that will be performed by the repair group and they will benefit from these training sessions. Many new products have modifications incorporated in various components and all these will require training in their repair and maintenance. Manufacturers and dealers also usually offer training to mechanics and electricians in major repairs or overhaul procedures. Some even offer operator training. The service group should ensure that all training offered by equipment builders and dealers is incorporated into the overall training program of the company.

## *MANUALS AND PUBLICATIONS*

The service manager should work closely with the repair manager in all phases of departmental work, especially training. Another item that requires close attention is the upkeep of the technical literature. With each piece of equipment, the manufacturer supplies parts books and service manuals. Depending on the particular type of equipment, there may also be repair manuals and operator's manuals available. Service manuals and operator's manuals are valuable as training aids in addition to their primary use as the basis for preventive maintenance, routine service, and proper operating techniques. Parts books and repair manuals are used primarily by the repair group.

At least one, but preferably two, complete sets of books must be kept by the maintenance department. If these reference books are not on file for any piece of machinery, contact the nearest dealer or the customer service department of the manufacturer. They can provide the books and manuals for your particular unit. All they need is the nameplate data from the unit. The cost of these publications is minimal and they are absolutely essential to the maintenance department. Without them, proper maintenance, operation, and repair are impossible.

Periodically, equipment manufacturers send out service bulletins to dealers and owners of their products. These bulletins may actually go by a number of different names, but they all serve the same purpose—to advise owners and dealers of changes. For example, a manufacturer may issue a change notice advising of a change in a part number or quantity. A service bulletin may advise that a repair part is obsolete and what part, or parts, can be substituted. Bulletins may also announce changes in service or maintenance procedures. Whatever changes are indicated, they must be entered into the appropriate publication before the bulletin is filed. If a modification kit is required, steps should be taken to order it for immediate use or as stock. Bulletins should not be permitted to accumulate for future action. Process them immediately on receipt. Make sure that the dealer and manufacturer have the company on their mailing lists. It is a good idea to arrange for these bulletins and change notices to be sent to the maintenance department manager or service manager. Do not have them sent to a person by name. When people change jobs, their mail is often thrown out or set aside and lost.

The service group should be in charge of seeing that the information received is distributed to those who need to know and that the various books and manuals are changed to show any new information. If other action is required by a bulletin, see that it is carried out.

## REPAIR DATA

The service group must work closely with the repair group to ensure that all data pertaining to equipment repair are properly collected and that costs are accurately assigned. Repair personnel must have their time charged to the particular unit they work on and all parts used must also be charged to that unit. When a work order is completed on the unit, this cost information will be transferred to the service group files. A brief description of the work performed will also be recorded. These data, together with those of routine maintenance and service, form a major part of the master plan. How this fits in with other parts of the plan will be explained later in the book.

## LUBE OIL MONITORING

If a system of lube oil monitoring is used by the maintenance department to monitor the internal condition of operating equipment, the service group should be in charge of its administration. Many companies are now using this new form of equipment management, which just now is beginning to evolve from its beginnings as simple lube oil analysis. The result of this new approach is a continuous monitoring of lubricating oils that goes beyond simply determining the condition of the lubricant at the time of sampling. Machinery is continually monitored not only to detect wear metals but also to observe the actual chemical condition of the lubricant and to monitor trends in the overall conditions of both oil and machine. By observing these trends, oil can be used until it is no longer in like-new condition, and necessary adjustments can be made to the equipment long before serious problems arise.

If this additional tool of maintenance management is used, the service group will logically be in charge of its proper operation and administration. In general, this means that routine scheduled sampling will take the place of scheduled oil and filter changes. In addition, the service group will be responsible for ensuring that maintenance procedures and adjustments are carried out when the monitoring system dictates the need.

A machinery-in-operation monitoring system is a welcome technology that permits the maintenance organization to add to its control over the equipment by having confidence in the internal condition of that equipment at all times. A monitoring system easily adapts to a basic master maintenance plan and requires little extra work on the part of service personnel. This state-of-the-art approach to maintenance management is especially needed in modern industries because of the complexity and high cost of mechanical equipment. This service has been available commercially for a number of years. Its value to industry is tremendous, and because of this, Chapters 9 through 11 are devoted to the subject of mechanical systems integrity management.

In this chapter we have described the major functions of a service group within a maintenance organization. The basics required for a start toward implementing a master maintenance plan have also been set forth. Beginning with Chapter 6, the full details of the master plan will be explained step by step. Because of the many and varied types of companies, the duties of the maintenance organization, and especially the service group, may differ somewhat. Basically, they include what has been outlined here.

# chapter 5

# Fuel and Lube Oil Management

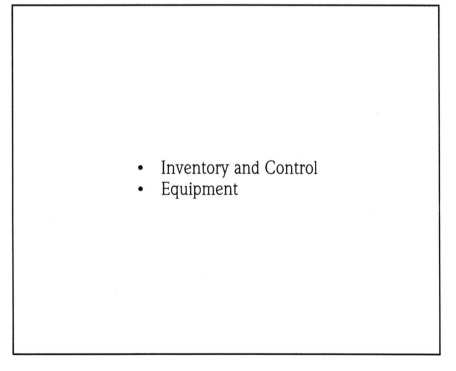

- Inventory and Control
- Equipment

## INVENTORY AND CONTROL

In companies where a large volume of fuel and lubricating oil is used, the maintenance organization should be responsible for overseeing its use and storage. The cost of these products can be a controlling factor in how effective an operation is being run. This cost factor will always be with us and will probably become more critical in the future. The price of petroleum products will doubtless increase, and the outlook is for little change in this trend.

Obviously, steps must be taken to set up a system of strict accountability for the use of all fuels and lubricants. The service group is the logical choice to administer the use of these products. Service should have the responsibility for dispatching fuel and lubricants to the equipment. Most fuels will be bulk stored in above- or belowground tanks. If the operation is big enough, lubricating and hydraulic oils will also be stored in large bulk tanks. Some special-purpose oils and greases will normally be stored in drums. There may be more than one storage depot. Fuels and lubricants may also be located at service stations, on lubrication service trucks, fuel trucks, remote platforms, and in or near repair shops. The service group will use all of these facilities to service equipment.

At the end of each month, all these depots and mobile service units must be inventoried. An example of an *inventory form* used for this purpose is shown in Fig. 5.1. One of these forms is to be filled out for each location that has fuel and/or lubricants. There is space available at the top of the form for the date and for indicating which location has been inventoried. There are columns for fuels, lube oils, automatic transmission fluid, and grease. There are a number of blank columns for writing in any types of petroleum products that might be added in the future, or for low-usage specialty products. When establishing the format of the form, all known products used should be included, as well as extra columns. It is not unusual to acquire a new piece of equipment and discover that it requires a lubricating oil different from any currently used by the company.

Not all locations or service trucks necessarily carry all the products listed on the inventory form. A fuel truck will only have one entry on its form, either gasoline or diesel. A lubrovan, or lube service truck, will generally carry some of every product. The important thing to keep in mind is to inventory each place separately and at the same time. Doing it this way will reduce the chance of omitting a location. A number of locations can be inventoried at the same time by different people. This reduces the total time required to complete the monthly inventory. If only one inventory sheet were to be used, it would be difficult to check if all locations were covered

## MONTHLY INVENTORY OF FUEL AND LUBRICANTS

SERVICE AREA OR TRUCK NO. _____          DATE _____

WAREHOUSE ☐        TANKER TRUCK _____

| DIESEL | GASOLINE | ROTELA T 30 | ROTELA 30 | SAE 40 | DONAX T 4 | AUTO TYPE A | 90 |
|---|---|---|---|---|---|---|---|
| GALS. | GALS. | GALS. | GALS. | GALS. | GALS. | GALS. | GALS. |
|  |  |  |  |  |  |  |  |
| TELUS 51 | GREASE |  |  |  |  |  |  |
| GALS. | LBS. |  |  |  |  |  |  |
|  |  |  |  |  |  |  |  |

_____
Authorized Signature

**Figure 5.1** Monthly inventory of fuel and lubricants.

and if not, which were left out. When the supervisor receives the completed inventory forms, he will know immediately whether or not all locations or mobile units have been inventoried. After using this system over an extended period, it will be relatively easy to determine if the inventory was properly taken. If an inventory sheet indicates a marked change from the norm, steps can be taken immediately to determine the cause. This method enables the supervisor to pinpoint the exact location of the discrepancy and who was responsible for inventorying that location. It makes it easy to check the inventory quickly and to spot theft quickly. These forms should be printed and bound in tablet form for ease in handling.

As soon as all inventory forms are received and checked by the service supervisor, the *monthly report of fuels and lubricants used* (Fig. 5.2) can be completed. This report should be sent to the maintenance manager and purchasing department. The accounting department may also wish to have this information. Copies of the report should be kept on file as a historical reference and as another data source to be used to measure the efficiency of the overall operation.

The report includes a column for each type of fuel and lubricant plus a blank space to be used for any additional product. When a maintenance organization sets up a system for administering the use and storage of fuels and oils, it first must establish exactly what products are in use. Once this is

# MONTHLY REPORT OF FUEL AND LUBRIANTS USED

To _____

From _____

Month of _____

| | DIESEL | GASOLINE | ROTELA T 30 | ROTELA 30 | SAE 40 | DONAX T 4 | AUTO TYPE A | 90 | TELUS 51 | GREASE |
|---|---|---|---|---|---|---|---|---|---|---|
| | GALS. | GALS. | GALS. | GALS. | GALS. | GALS | GALS | GALS. | GALS. | LBS. |
| PREVIOUS BALANCE | | | | | | | | | | |
| RECEIVED THIS MONTH | | | | | | | | | | |
| TOTAL | | | | | | | | | | |
| BALANCE — MONTH END | | | | | | | | | | |
| AMT. USED THIS MONTH | | | | | | | | | | |

**Figure 5.2**  Monthly report of fuel and lubricants used.

done, the forms for inventory and reporting can be designed around these products.

The totals received each month can be obtained from the purchasing department if the maintenance department does not have responsibility for receiving these products. The balance at month's end is the total of all the monthly inventory sheets for each product. Subtracting this from the previous balance plus amounts received will give the total usage of each product for the month.

These total monthly usage figures can be compared to the totals dispatched to the operating units and other miscellaneous users. There will obviously be some differences caused by inaccuracies in data recording, especially in amounts dispatched to the various units. The minor variances will tend to accumulate over the month. If there is a noticeable discrepancy, after taking normal variances into account, steps can be taken to determine the cause. It may take some time to add up all the monthly totals dispatched to each unit, but it will prove worthwhile if management suspects something is not right in the system. If the dispatch data are fed into a computer, the job is much easier. The objective is to investigate suspected discrepancies as soon as the monthly inventory is completed. Under no circumstances should this be postponed until year's end. By that time, any problem discovered will be much greater and more difficult to isolate and correct. Using this system of checks and balances will help establish firm control over the use of petroleum products and ultimately, that all-important factor, cost.

The monthly report forms can also be made up in tablets for ease in handling and storage. Since it is the information that is most important, it is not necessary to type these reports. Slip carbons into the tablet and fill out each report by hand. This will save time and reduce the chance of errors being introduced.

## EQUIPMENT

Larger operations in which equipment is scattered over a large area should make use of mobile lube service trucks, commonly called lubrovans. As an alternative, a flatbed trailer is sometimes used together with a truck or tractor to haul it. Either configuration can provide the same services and include the same equipment. In general, they can be set up to carry a variety of services and the layout may vary depending on the type of equipment to be serviced. The platform of a lubrovan may be open, semienclosed, or fully enclosed. The deciding factor will be the user's preference and cost. These units can be purchased ready for use or the components can be bought separately and mounted on existing equipment. Lubrovans, or lube trailers, can be equipped to include any, or all, of the following:

1. Diesel fuel
2. Gasoline
3. Lubricating oils
4. Hydraulic oils and grease

5.  Compressed air
6.  High- and low-pressure water

They are also usually equipped with space to carry filters, tools, antifreeze, anticorrosion additives, and distilled water flasks.

All delivery pumps are operated by compressed air. A tank-mounted compressor can be driven by either a small air-cooled engine or from the power takeoff on the truck. The former is usually preferred. Diesel fuel tanks can be as large as 800-gallon capacity. Generally, they are in the 500-gallon range. There should be a filtering system in the fuel delivery line to remove water and sediment. Lubrovans rarely service units that use gasoline and so do not normally carry that fuel. In some cases, a 5- or 10-gallon portable tank is carried for emergency use or for cleaning purposes. In this case, gasoline would be dispatched by hand. Oils and grease are carried in 55-gallon drums. The grease drum is usually designated by weight, the most common size being the 400-pound drum, which looks like a 55-gallon drum. Drums are mounted on end with the bungs up and the bottoms nested in retaining rings. The drums are secured in place by rods that grip the rings and drum tops. The method of tightening down varies. An air-actuated pump and suction tube is set on the drum head with the tube extending down through the bung opening. Some manufacturers supply special drum covers that serve to clamp the drum in place and provide a more rigid pump mount. These air pumps are available in different capacities and qualities. As a rule of thumb, the better the quality, the lower the maintenance cost on both pump and tube. Regular pressure pumps will handle most oils. High-pressure pumps are required for greases and extra-heavy oils.

From each pump a hose leads to a reel at the rear of the platform. Each reel is equipped with a hose and an applicator. These applicators must have counters on them, graduated in gallons or liters. The exception is the grease applicator.

There should also be reel assemblies for compressed air and water. The air is used primarily for tire inflation, high-pressure water for cleaning equipment, and low-pressure water for use in cooling systems.

Mobile service units, on a smaller scale, can be made up for use on pickup trucks. These are usually for limited types of service. Lube service for inside stationary equipment can be provided from small hand-pulled or electric-powered wagons. These would carry hand pumps and graduated pitchers for measuring amounts dispatched to each piece of equipment.

Other methods of providing fuel and lubricants are from fixed lube and fuel platforms and from service stations. Service platforms are usually equipped to provide only fuel and lubricating oils. They may be quite primitive, consisting of a raised rack for holding oil drums. The drums are placed on their sides with faucets in the bungs. Nearby is usually found a small raised fuel tank with a gravity-feed arrangement. These remote platforms are not equipped for greasing except for a hand gun and usually do not carry a full line of oils. Unless absolutely necessary, the use of remote platforms is not recommended. The major drawback is that they are not secure, and it is difficult, if not impossible, to control the dispersal of fuel and oils unless someone is stationed on them at all times.

The best method is to provide all field service from lubrovans or lube trailers operated by service personnel. Service technicians are trained in the proper methods of giving lube service and accounting for the quantities dispatched. Normally, equipment operators should not be allowed to do their own servicing. If they do, there is a very good chance that quantities used will not be recorded or reported and an incorrect lubricant will be used. For many operators, oil is oil. Whether coconut oil or 140-weight gear oil, it is all the same to some. Mixing oil in engines can have disastrous consequences that go beyond just lubricating characteristics. The additives in some oils mixed with those in others can form chemical reactions. It is best to leave lube service to people trained for that job.

Service stations should provide the full range of services required by mobile equipment. In many ways, they resemble the service stations located throughout the country that cater to automobile owners. The major difference is in the size. These stations must have the space and facilities available to handle autos, trucks, buses, tractors, and any other type of large off-road machinery that can be driven in for service.

Service stations should be located near repair facilities to take advantage of the services available and to save time. For example, it simplifies matters if, after repair work, a unit can be moved nearby for full lube service prior to putting it back to work. Conversely, if during lube service it is discovered that repair work or mechanical adjustment is necessary, the facilities are near at hand.

In addition to having all the petroleum products available and the equipment to deliver them, the station should be equipped with lifts and pits for greasing and for oil changing. There should also be a washing area near the station. Stream cleaning units should be used. Cleaning machinery is a vital part of service. Equipment should be cleaned prior to lube servicing or repair. It is much easier and more efficient to work on a clean machine. Cleaning is also a good way to reduce losses. It is easier to observe potentially dangerous leaks if the unit is clean and not covered with a paste of dirt and oil. A major problem may be avoided by correcting a source of leakage. Fire is another hazard associated with dirty equipment. This is more of a problem in industries such as pulpwood, forestry, agriculture, and construction than in cleaner industries, but in all cases, dirty machinery is a safety problem. The disastrous consequence of fire must be avoided by all enterprises, whether considered clean or not. It is simply common sense. A spark or carelessly discarded cigarette can ignite an oily machine, causing loss of the unit and perhaps the building or vessel in which it is located. Cleaning is an inexpensive way of reducing fire hazards and maintaining machinery in good working order.

Fuel tank trucks are used in some instances, especially where a large amount of equipment is spread over a wide area. These will lessen the work load of the lubrovans in field operations and speed up fuel deliveries.

Some operations with stationary units have their own separate fuel supplies. Some engines run on natural gas. Meters are installed in the lines to record the amount being used. Whatever the system, steps must be taken to ensure that an adequate amount of fuel is supplied and that the amount consumed is accurately recorded.

In summation, the service group must have control over the storage of all fuels and lubricants as well as their use. This involves inventory control, dispatch and use of petroleum products, equipment, facilities, and the personnel required to carry out this mission. This holds true regardless of the size or type of industry. The equipment may vary and the service requirements differ, but the basic approach remains the same.

# chapter 6

# Laying the Foundation

- Status Boards
- Periodic Service Sheets

In Chapter 4 we discussed in detail some of the steps that are necessary to set up an effective service system and master plan. The equipment inventory list, equipment file folders, and maintenance history cards have all been explained. Now that these items have been completed, what comes next?

There are two additional basic aids that must be set up to complete laying the foundation for the preventive maintenance master plan. The first is to set up a visual display showing the current status of all equipment. The second task is to review the routine service and maintenance requirements for each piece of equipment and combine these into a common system for providing service. This is done by making up periodic service sheets, or instructions, for each unit or type of unit.

Together, these five steps will require the most time and effort on the part of the service group. It will take a lot of digging to obtain all the information available on each machine and to compile the inventory list and set up files, displays, and service sheets. Once it is done, it should never have to be done again. The only changes will occur when new equipment is added or old equipment is removed from service. To maintain equipment properly, those in charge must know what they have under their jurisdiction and what is required in the way of preventive maintenance. Once this is thoroughly established, the rest is easy. It is amazing how many maintenance organizations are unfamiliar with their equipment and are disorganized to the point where they are not certain how to go about providing proper routine service. It should be clear by now that what is being set forth in this book is nothing more than detailed organization and control.

## STATUS BOARDS

One of the major problems in any preventive maintenance system is not knowing the current status of equipment and if service is being provided on schedule. The supervisor and the service technicians know because they are working with the machinery on a day-to-day basis. It is management that is usually unsure of exactly what is going on and often does not have enough background information to ask the right questions and get good answers. The more units involved, the more difficult it is to keep track of everything. The service supervisor and service personnel may let something get by them. The best-kept files alone will not solve this problem.

The solution is to get as much information as possible out in full view for everyone to see. This not only helps the managers immediately to get a complete picture of the current situation, it also helps the service group to stay abreast of the service needs of each unit. One type of visual display that

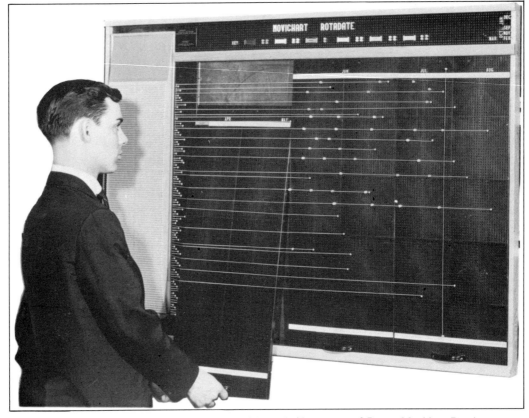

**Figure 6.1**  Visual control board. (Courtesy of Pryor Marking Products, Inc.)

has proved to be very effective is the *visual control board* (Fig. 6.1). These are manufactured under different trade names, but they are essentially the same in their make up and function. This type of visual display is used for all types of businesses and systems. It has applications in sales, maintenance, inventory control, and production control, to name just a few. The interesting thing about these boards is that regardless of where they are made, the word *control* is used somewhere in their description. This is what their main function is: to allow easy control over an operation without having to waste time going through endless files hunting for data. The entire operation is in full view, giving an instant picture of the status of each unit.

A typical board is capable of listing 100 units. A visible card pocket panel is mounted on the left side of the board. A 5 by 7 inch index card (Fig. 6.2) can be inserted into each of the flip-up card pockets, which are numbered 1 through 100. The bottom of each card is visible and is in line with a tracking peg located to the right of the card. One index card is completed for each unit. Since the bottom line of the card is all that shows, this is where the descriptive data are printed. These data should be in the same format as on the master equipment inventory list. From left to right are listed the following: company number, make, or name description, model, and serial number.

These cards should be arranged in exactly the same order as the master equipment inventory list, with the lowest number at the top. If a second or third board is required, they continue in order with the lowest number at the

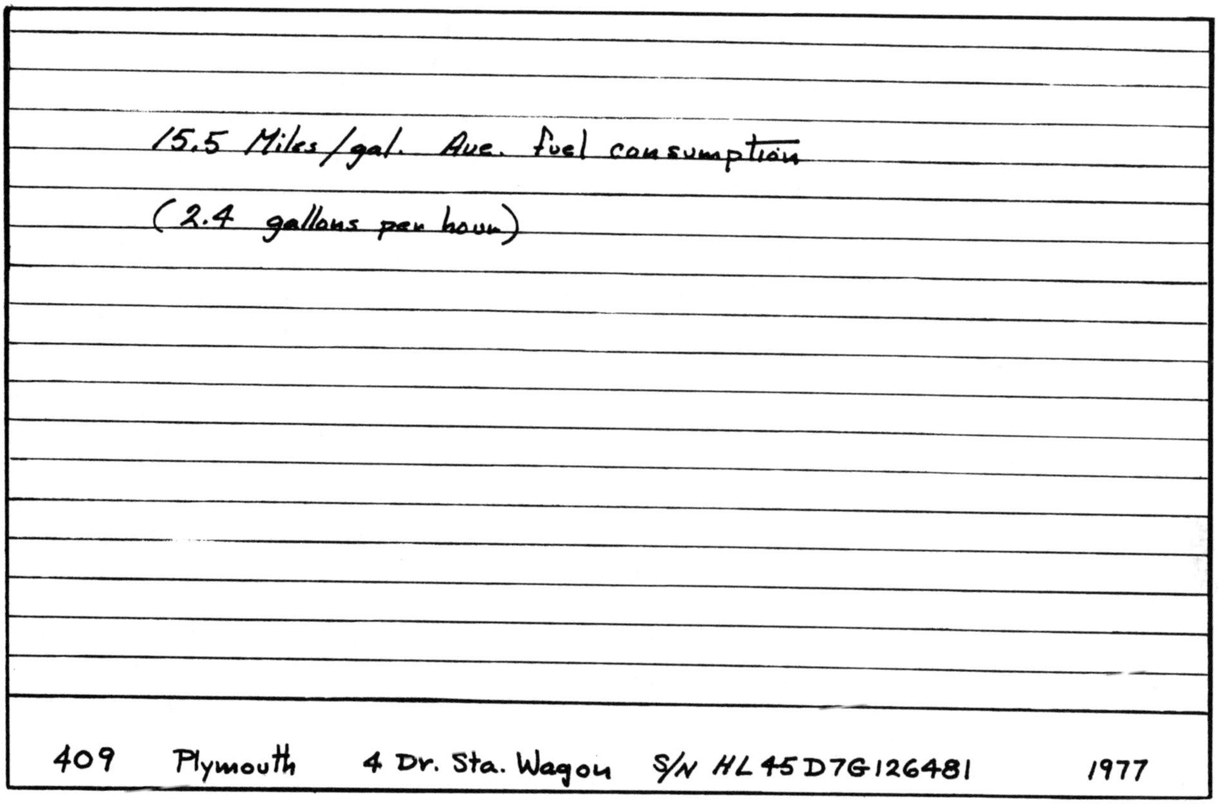

*15.5 Miles/gal. Ave. fuel consumption*

*(2.4 gallons per hour)*

*409   Plymouth   4 Dr. Sta. Wagon   S/N HL45D7G126481   1977*

**Figure 6.2**  Control Board Index Card

top increasing toward the bottom. Not only should the order of company numbers be the same, but the list format should be the same. Group headings should be included in the card listing just as they appear on the master equipment inventory list. Thus there will be some cards that have only a group heading: for example, "Generator Sets." There are two reasons for this. First, everyone involved with the system will find it easier to function when everything is in the same order and format. Using the same setup for equipment lists, status boards, and files will impart a sense of order when dealing with a large amount of diversified equipment. Second, group headings serve to break up the list and make it easier to read and locate a particular unit. It becomes confusing if the board contains 100 listed units with nothing to relieve the eye. The group headings act as a quick index to particular units listed on the board.

As on the master equipment inventory list, there should be a number of blank spaces left at the bottom of each group. This allows space for adding more units in the future. It is easier to insert a new unit card in an empty space than to move all the subsequent cards down one space. Imagine the extra work rearranging boards if a new card has to be inserted somewhere at the top of the first board and there is no space available. Not only would all the cards need to be rearranged, but the tracking pegs would also have to be repositioned to suit the new number they represent. This could take hours and introduce errors in the positions of the status pegs.

Remember, this is a visual display, so the boards must be easily readable and equally easy to maintain. So use the group headings and allow plenty of extra space. If there are 195 different pieces of equipment to keep track of, use three boards.

Old company numbers, those on units removed from service, can be reassigned to new units. However, it is a good policy to wait awhile before doing so to avoid confusion when studying operating and ownership cost histories. Wait until a new calendar or fiscal year begins before reusing an old number.

It is not necessary to put all the information from the master equipment inventory list on the bottom line of the card. Only the data shown in Fig. 6.2 are usually listed. Any additional information that is considered necessary can be noted on the portion of the card that is hidden from view. It can be read easily by lifting the card above it. The bottom line of the card is simply an identification line.

The largest section of the board is located to the right of the row of tracking pegs and is covered with drilled holes to accept the pegs. From left to right, there are 200 holes. This may vary with the type of board. The peg section of the board is marked off with horizontal and vertical lines every 10 holes to make it easier to keep tracking pegs at the proper level and station.

Boards are set up to indicate the status of the equipment based on hours, days, or miles. Any one of these can be used to indicate when periodic service is to be given to the equipment. The exception to this is daily service given to all units in operation and is not posted.

Status boards are supplied with loose colored pegs and colored strings. The colored pegs can be used to indicate changes in status of various units. The user can make up a color index showing what each indicates. For example, a red peg alongside a tracking peg might indicate that the unit is out of service and on standby status; a yellow peg might indicate that the unit is in the shop for repairs; and a blue peg might show that the unit has just received its scheduled service.

The colored strings are set up vertically on the board and generally are used to indicate the periodic service intervals. Different colors are used to mark service at 100, 250, 500, and 1000 hours. If hours are used as the basis for periodic service, the board should be set up on a 1000-hour scale. This can vary up or down, depending on the type of equipment used and the service requirements. If 1000 hours are used, each hole will represent 5 hours of work. After 1000 hours, the cycle is repeated with the tracking peg set back at the left-hand starting point.

All this is provided so that management and supervisory personnel can see at a glance the status of equipment service and routine maintenance. It also provides a quick check on how one unit compares with others. It shows which units are working, on standby, or down for repairs. No one has to ask what is going on or needs to check through file cabinets. It is all there, on the wall, in plain view. The service group supervisor should have direct responsibility for keeping these boards current. By doing this every day, he becomes intimate with each and every unit. It is nearly impossible to have a neglected unit with this type of system.

Since most manufacturers' recommended periodic service and maintenance are based on hours, this is normally used as the common basis for servicing equipment. However, in the case of cars, trucks, and buses, mileage may be used as the controlling factor for providing periodic service.

There are other types of equipment that may have service based on weekly, or monthly, intervals. In general, the various units should have all their servicing performed on a standard basis for the sake of simplicity. There are always exceptions to the rule, so the methods used will vary with individual operations. For example, a company with a large group of off-road equipment plus a fleet of trucks might use two systems. One board would be set up based on hours for the off-road equipment and another set up based on miles, or kilometers, for the trucks.

When providing periodic maintenance and service, the main thing to keep in mind is that it must be done on time. Service given on a hit-or-miss basis, or when convenient, is not good enough. A method must be used that provides an accurate indication of elapsed working time on the equipment. The key word is *work*. A unit may not be fully employed during the working day for any number of reasons. To simply count up the hours in a normal working day and apply this figure to every unit not on standby, or undergoing repair, would be inaccurate and probably lead to unnecessary and costly excess servicing.

In the case of off-road equipment and some stationary equipment, time can be read from hour meters mounted on the machine. Trucks, cars, and buses are equipped with odometers that record miles or kilometers. These instruments must be kept in good working order. In the case of equipment that uses internal combustion engines, a more accurate indication of work done is to measure the amount of fuel consumed. This information is easily available since it is recorded each time a unit is refueled. Service based on fuel usage is best suited to maintaining equipment. The harder a unit works, the more fuel it consumes, and the service intervals automatically become shorter. Conversely, if a unit is being used for light duty, or something less than full capacity, the service interval will be longer because the amount of fuel used per time period will be lower than normal. By studying manufacturers' literature, an average fuel consumption rate can be obtained in gallons per hour. This is available for all mobile and stationary equipment. By applying an average fuel rate to the amount of fuel used during the day, service personnel can calculate the hours worked by a unit and use this figure to move the appropriate tracking peg on the status board.

This system does not work with all types of machinery. Some units should have service based on elapsed time. This means taking into account the number of normal working hours in a day and applying this figure to the working unit. Equipment whose hours or mileage is read from meters must be checked regularly to ensure that these instruments are in good working order. Unfortunately, it is not uncommon to have inoperative or inaccurate gauges, and it is difficult to give precise service to those units. Road vehicles can be serviced on the basis of fuel consumed, but it is more difficult to arrive at a fuel rate for them. It can be done if enough is known about the type of work done and the conditions in which each vehicle will be working. Past experience may permit a close estimate of the mileage run for a full working day, and from this an average speed can be calculated. Similarly, an average full consumption rate in miles per gallon can be established. By dividing the average speed by the average miles per gallon, an average fuel rate in gallons per hour can be obtained. As in the case of off-road equipment, dividing the amount of fuel dispatched each day by the unit's fuel rate will give the calculated hours worked. This figure determines how far the tracking peg will be moved on the board for that day.

A final thought concerning servicing on the basis of elapsed time in hours should be mentioned before ending the discussion. Most off-road and stationary equipment are provided with hour meters. These record off the engine and give a cumulative total of running time. Obviously, this can be noted each time the unit is fueled and applied to the status board, after subtracting the reading from the preceding fueling. There are some drawbacks to this, the most common being the probability of broken or inaccurate meters. It also adds one more chore for the service technicians, that of recording times. This increases the chance of error, especially where there are large amounts of equipment to be serviced each day. The hour meter records the time that the engine has run but does not take into account variations in the degree of work. Work-load variations result in changes in rpm and in the amount of strain put on the engine, both causing changes in the rate at which fuel is used. Thus the hour meter is not as accurate an indicator of service intervals as is the amount of fuel consumed. This is not to say that a good workable servicing program should not be based on hours. Indeed, it usually can be made to work quite well. This method is not as accurate or foolproof as one based on fuel usage.

Whether or not a fuel rate, in gallons per hour, is used to determine accumulated time on a status board is for the maintenance department to establish. In any event, the fuel rate should be established and noted on the index card of each unit. This way, if a working unit has a broken hour meter, its working hours can be quickly calculated each day and servicing can proceed without gaps.

The most important point to remember is that periodic service and maintenance must be given on a regular established basis, not in a random manner. Over the long run, it matters less whether the service is based on hours, days, mileage, or fuel usage than if there is strict control over the interval used for triggering periodic service and maintenance. Finally, the service group must ensure that service is performed in accordance with manufacturers' recommendations.

## PERIODIC SERVICE SHEETS

Periodic service sheets are made so that everyone involved will understand exactly what is required in order to provide proper service and maintenance to each piece of machinery. This information is available in equipment service, maintenance, and operator's manuals. Unfortunately, these publications often are not properly disseminated by maintenance departments and few people get to read them. Information filed away in an office bookcase is of very little use. Service supervisors and technicians should have them on hand for study and as a ready reference. Unless this information is made available to the personnel who need it, periodic maintenance will degenerate to refueling and occasional lube service.

Periodic service sheets should be made for each model or type of equipment in the company's inventory. For example, pickup trucks can be combined so they are covered with one common set of periodic service sheets. Tractors and other mobile or stationary equipment can be covered by make and model. If there are 10 ore haulers of the same make and model, they will use the same periodic service sheets. With some machinery, similar types may use a common set of sheets. It all depends on how similar are the maintenance instructions issued by the different manufacturers.

Using the master equipment inventory list as a guide, determine if there are manuals on hand for all models listed. If not, order them from the dealer or the manufacturer. Begin by studying the sections dealing with periodic maintenance. These will explain in detail how to go about giving service and the recommended intervals for each type of service. There will also be information about the type of lubricants to use, cooling system additives recommended, adjustments, and operating limits. Usually, these manuals contain a short section that summarizes the service to be given by increments of time or mileage. Use these sections to set up the periodic service sheets for that model or type.

Service sheets will vary in complexity. The sheets covering a large reduction gear or pump engine will be very simple, but those for a large tractor or mobile crane will be more complex and detailed. This is due to the fact that the latter types incorporate many different components, each requiring special service.

To illustrate how a set of periodic service sheets are set up, we will examine the requirements of a typical large tractor. This could be either a construction or an agricultural unit and might be described as an articulated tractor powered by a 250-horsepower diesel engine. For this example, we will also add that it is equipped with an automatic transmission and a drop-type gearbox. It is equipped with a hydraulic system separate from the transmission. The service requirements of such a unit are varied and well suited to serve as a model for explaining the process of making periodic service sheets.

The first step is to determine what must be done on a daily basis, other than fueling the unit. The daily sheets are the only ones that will be separated from the routine followed when working with the others. These sheets should be collected in book form with all the other daily service sheets. Loose-leaf binders are ideal for this purpose. These daily service books should be distributed to every service station, lubrovan, shop, area, or person that will be involved with machinery servicing. In addition, all supervisors who are in any way involved with equipment should have a copy. Old hands, who service machinery every day, will not need to refer to these pages very often, but they do provide a quick reference. They are invaluable to less experienced or new employees. The objective is to get the detailed instructions on paper so that this vital information is always available and consistant. Information stored in the heads of a few old employees is neither reliable nor safe. If key people are lost, so will be their knowledge, and serious service problems will result.

Figure 6.3 shows what a daily sheet for this typical tractor would include and how it is set up. The make and model of the tractor are shown at the top of the page. In this particular example, the company bases its daily service on an average working time of 10 hours. However, the indicated service is done each day rather than at a given interval of hours. If the unit

```
              (MAKE OF) TRACTOR                    (MODEL)


                     DAILY (10-HOUR) SERVICE



                          LUBRICATION

Articulation Joints (2)                              MPG


                            SERVICE

1. Check engine oil level.   SAE 30

2. Check air filter element. If red, clean air filter.

3. Check radiator coolant level. Blow clean radiator/grill.

4. Check level of oil in hydraulic tank.   SAE 30

5. Check transmission oil level.    ATF A

6. Check transfer case oil level.   EP 90

7. Visually check the following:
          Batteries and terminals        Belts
          Tires                          Gauges
          Lights                         Controls
          Hoses
```

**Figure 6.3**  Periodic service sheet: daily.

were to work under extreme conditions, 24 hours per day, it might be advantageous to use the 10-hour schedule.

This sheet is divided into two groupings, one for lubrication and one for service. For this example, lubrication is taken to mean components requiring grease. MGP stands for "multipurpose grease." The sheet shows that two grease fittings on the articulation joints require daily greasing. Under the service section are listed the various items that must be checked daily. Coolants and oils must be kept at the proper levels. To the right of the numbered listings are shown the type of oil to be used, where appropriate. The last item is a checklist of those things that require daily attention.

Figures 6.4 through 6.8 are service sheets for 50-, 250-, 750-, 1000-, and 1500-hour intervals for a typical tractor. For this particular unit, there are three areas that require lubrication on a 50-hour, or weekly, basis (Fig. 6.4). They are listed along with the number of grease points for each. The format of this sheet differs from the daily sheet in two ways. At the top right corner is a space to write in the company number of the unit being serviced, and at the bottom are spaces for the service technician's signature and date. This format is common to all service sheets, with the exception of the daily sheets.

When the service clerk, or supervisor, sees that a particular unit has completed 50 hours of work, he takes a 50-hour service sheet for that make and model and writes the unit number in the space at the upper right corner. The sheet is then given to the technicians who will service the unit. When the service has been completed, the service sheet is signed, dated, and turned in to the service supervisor at day's end. The supervisor now knows the service is done and can so indicate on the status board. Daily service continues until 100 hours are accumulated by the tractor. At that time, another 50-hour sheet is issued. This routine continues until the unit has worked 250 hours. At this point, a special service sheet for 250 hours is issued (Fig. 6.5). Routine daily service is still given, but now there are additional areas that require lubrication and special servicing.

To keep the system as simple as possible, a 50-hour sheet will not be issued along with the 250-hour sheet. The three component parts that are to be lubricated every 50 hours are included as the first three items under the lubrication section of the 250-hour sheet. Compare Figs. 6.4 and 6.5. The same would hold true if there had been any service items for 50 hours. In this example, there are none. The new items to attend to at 250 hours are 4 through 7 under lubrication and the three listed under service.

This sheet is issued to the service technician in the same manner as was the 50-hour sheet. There is one difference—this sheet lists an engine oil and filter change. So, in addition to issuing the service sheet, the service supervisor will include the proper filter for that tractor engine. In filling out a material requisition for the filter, the supervisor must list the unit's company number so that this filter will be charged to the proper tractor. The service technician, provided with the service sheet and filter, can now give the proper service for 250 hours. When finished, the technician turns in the signed and dated sheet.

Daily service continues and at 300, 350, 400, and 450 hours, a 50-hour sheet is issued for the tractor. When the unit has accumulated 500 hours, another 250-hour sheet is issued. This is so because, in this example, the

```
                                    Unit No.




          (MAKE OF) TRACTOR                    (MODEL)



              50 HOUR SERVICE (WEEKLY)

                   LUBRICATION

     1. Oscillation supports on front axle (2)      MPG

     2. Mid-ship support bearing (1)                MPG

     3. Steering cylinder pivots (4)                MPG

















          Date service completed:          Signed:
```

**Figure 6.4**  Periodic service sheet: weekly.

```
                                              Unit No.


   (MAKE OF) TRACTOR                    (MODEL)



                  250 HOUR SERVICE



                    LUBRICATION

   1. Oscillation supports on front axle (2)      MPG

   2. Mid-ship support bearing (1)                MPG

   3. Steering cylinder pivots (4)                MPG

   4. Engine fan drive pulley bearing (1)         MPG

   5. U-Joints and shafts (10)                    MPG

   6. Axles and bearings (4)                       MPG

   7. Valve handle bushings (1 to 6)              MPG


                     SERVICE

   1. Change engine oil and filter.  SAE 30

   2. Drain water and sediment from fuel tank.

   3. Check oil level in front/rear differentials. SAE 30




   Date service completed:          Signed:
```

**Figure 6.5**  Periodic service sheet: 250-hour.

manufacturer has not specified any special service requirements at 500-hour intervals.

The routine continues with 50-hour sheets being issued at 550, 600, 650, and 700 hours. At every 750 hours, the manufacturer has specified additional special service to be performed, so a 250-hour sheet cannot be used. The 750-hour sheet (Fig. 6.6) includes all the lubrication and services specified on the 250-hour sheet plus the 50-hour sheet.

The seven items listed under lubrication are the same as on the 250-hour sheet. In fact, these seven appear unchanged on all the service sheets, starting at 250 hours. At 750 hours there are two additional areas of service that need attention. These are shown as items 4 and 5. The supervisor now draws engine, transmission, and hydraulic oil filters from stock and charges them to the unit receiving this service. The supervisor then issues the three filters, along with the service sheet, to the appropriate service technician.

Work continues with daily and 50-hour service being performed until the tractor accumulates 1000 hours. Since there are requirements unique to 1000-hour intervals, a special sheet is used (Fig. 6.7). Lubrication instructions are unchanged, but the last four items under service are different.

Item 3 calls for changing, instead of checking, the oil in the differentials. Item 4 calls for changing the fuel filter element. It should be noted that fuel filters must be changed whenever they become dirty enough to cause problems with fuel flow. This varies widely with operating conditions and quality of fuel. Since the manufacturer has seen fit to list only a routine change at 1000 hours, the owner has gone along with the schedule with the periodic service sheets. However, the owner should keep in mind that conditions may require more frequent and unscheduled changes.

Item 5 calls for servicing the cooling system. This is a very important requirement, particularly for modern engines with pressurized systems. An entire chapter could be devoted to the subject, but it is outside the scope of this book. It is enough to say that it is essential that the service group study the section of the service manual dealing with this topic and service the cooling system according to instructions. Item 6 is self-explanatory.

After the 1000-hour service mark, the tracking peg on the status board is reset at the starting point. As most equipment manufacturers do not extend their recommended routine maintenance cycles beyond 1000 hours, this fits in well with the time scale set up on the status board. There are instances, however, when the manufacturer calls for a particular service to be done at greater intervals, as is the case with our tractor. Figure 6.8 illustrates the special service to be done at 1500 hours.

Between 1000 and 1500 hours, the normal daily and 50-hour service continues until 1250 hours. At this point, a 250-hour service sheet and filter are issued. Afterward, the normal routine is followed until 1500 hours have been accumulated. This sheet lists the same lubrication items as the others, and the service is similar to that of the 250- and 750-hour sheets. It is the same as the 250-hour sheet down through service item 3 and the same as the 750-hour sheet through service item 4. The difference is that items 5 and 6 call for cleaning the hydraulic tank and strainer plus changing hydraulic oil and the filter element. Refer to Figs. 6.5 and 6.6. This can be managed easily on a 1000-hour scale board by flagging the tracking peg hole with a

Unit No.


(MAKE OF) TRACTOR                          (MODEL)


750 HOUR SERVICE


LUBRICATION


1. Oscillation supports on front axle (2)      MPG

2. Mid-ship support bearing (1)                MPG

3. Steering cylinder pivots (4)                MPG

4. Engine fan drive pulley bearing (1)         MPG

5. U-Joints and shafts (10)                    MPG

6. Axles and bearings (4)                      MPG

7. Valve handle bushings (1 to 6)              MPG

SERVICE


1. Change engine oil and filter.  SAE 30

2. Drain water and sediment from fuel tank.

3. Check oil level in front/rear differentials. SAE 30

4. Change transmission oil and filter element. ATF A

5. Change hydraulic filter element.


Date service completed:              Signed:

**Figure 6.6**  Periodic service sheet: 750-hour.

<u>Unit No.</u>

<u>(MAKE OF) TRACTOR</u>                                    <u>(MODEL)</u>

1000 HOUR SERVICE

LUBRICATION

1. Oscillation supports on front axle (2)          MPG

2. Mid-ship support bearing (1)                     MPG

3. Steering cylinder pivots (4)                     MPG

4. Engine fan drive pulley bearing (1)              MPG

5. U-Joints and shafts (10)                         MPG

6. Axles and bearings (4)                           MPG

7. Valve handle bushings (1 to 6)                   MPG

SERVICE

1. Change engine oil and filter.  SAE 30

2. Drain water and sediment from fuel tank.

3. Change oil in front/rear differentials. SAE 30

4. Change fuel filter element.

5. Flush radiator and replace coolant and inhibitor.

6. Change oil in transfer case. EP 90

Date service completed:                    Signed:

**Figure 6.7**   Periodic service sheet: 1000-hour.

```
                                                      Unit No.


    (MAKE OF) TRACTOR                        (MODEL)



                    1500 HOUR SERVICE



                        LUBRICATION


    1. Oscillation supports on front axle (2)        MPG

    2. Mid-ship support bearing (1)                  MPG

    3. Steering cylinder pivots (4)                  MPG

    4. Engine fan drive pulley bearing (1)           MPG

    5. U-Joints and shafts (10)                      MPG

    6. Axles and bearings (4)                         MPG

    7. Valve handle bushings (1 to 6)                MPG

                          SERVICE


    1. Change engine oil and filter.  SAE 30

    2. Drain water and sediment from fuel tank.

    3. Check oil level in front/rear differentials. SAE 30

    4. Change transmission oil and filter element. ATF A

    5. Clean hydraulic tank and strainer.

    6. Change hydraulic oil and filter element. SAE 30


  Date service completed:            Signed:
```

**Figure 6.8**  Periodic service sheet: 1500-hour.

colored peg. This signifies that as the second 500-hour board cycle is reached, it will be treated as the 1500-hour service point.

To avoid confusion and not disrupt the normal pattern of routine service, the tracking peg must be returned to the starting point after completing the 1500-hour service. If this were not done, the 750-hour service would be done 250 hours later. This becomes wasteful and costly since that service was included in the 1500-hour service.

It is tempting to add the hydraulic oil change and tank and strainer cleaning back on the 1000-hour sheet, keep all service within a 1000-hour cycle, and eliminate the 1500-hour sheet. This can be done with some types of lubrication and service without causing much overmaintenance. In this case, the cost of the hydraulic oil and filter plus service time makes it uneconomical to reduce the service interval by 500 hours.

As stated earlier in this chapter, it is going to take time and a lot of effort to make up a set of service sheets for each make or model of machine. After studying the service and maintenance manuals, use the service summary as a guide and make a draft of the service sheets. Next, compare the sheets with the service manual and make any needed corrections. Write them so that only one sheet will be used for each service interval. Do not use several sheets covering all the intervals up to, and including, the current one. For example, the 250-hour sheet should include everything on the 50-hour sheet. Some rewrite and editing will be needed before all sheets are correct. It will also take a while before the service supervisor and clerks become thoroughly familiar with the use of the status boards. By the end of the first month of full operation, they will become adept in the use of both service sheets and status boards.

Except for the daily sheets, the service sheets are to be provided with space to note the company number of the unit being serviced and spaces for the service technician to sign and date the sheet when the service has been completed. Keep sufficient clean copies of all sheets on hand and file them by make, model, or type and by service interval in handy file cabinets. Daily sheets are to be compiled in binders and distributed to those people who need this information at hand. Daily sheets can also be posted on mobile equipment and on or near stationary machinery for quick reference. Sheets that are returned signed and dated do not have to be kept once it is acknowledged that the work has been done, usually by placing a colored peg or indicator on the board.

Keep in mind that nothing ever runs completely trouble free and a maintenance system is no exception. There will be times when an engine oil change may be required because of some unforeseen circumstance, such as fuel dilution. Water or dirt may get into fluid reservoirs and accidents can cause unscheduled repairs and servicing. When these things occur, adjustments must be made on the status boards and to the service schedule. A degree of flexibility is required but it need not interfere with the overall scheduled periodic service and maintenance.

As a further illustration of how to set up periodic service sheets, Figs. 6.9 through 6.12 demonstrate how one company chose to handle the servicing of light vehicles. They had approximately 20 vehicles, both two- and four-wheel drive. There were only three or four separate makes involved so they decided to combine them in one set of service sheets. Some items listed

<u>CARS</u> = <u>PICKUPS</u> = <u>JEEPS</u> = <u>BRONCOS</u>

DAILY 10 HOUR SERVICE

SERVICE

1. Check engine oil level.

2. Check coolant level in radiator.

3. Check transmission and transfer case oil levels.

4. Check level of brake fluid.

5. Check power steering oil level.

6. Check battery and cables.

7. Check windshield washer reservoir and nozzles.

8. Check the following for proper operation:

| | | | |
|---|---|---|---|
| Tires | Hoses | Clutch | Windows |
| Lights | Belts | Muffler | Master Cylinder |
| Brakes | Gauges | Windshield Wipers | |
| Controls | Horn | Mirrors | |

**Figure 6.9**  Periodic service sheet: daily.

<u>Unit No.</u>

<u>CARS = PICKUPS = JEEPS = BRONCOS</u>

50 HOUR SERVICE (WEEKLY)

LUBRICATION

1. Complete chassis lubrication (see chart).

2. Lubricate accelerator linkage.

SERVICE

1. Check oil levels in differential(s).

Date service completed:                              Signed:

**Figure 6.10**  Periodic service sheet: weekly.

<u>Unit No.</u>

## <u>CARS</u> = <u>PICKUPS</u> = <u>JEEPS</u> = <u>BRONCOS</u>

### 250 HOUR SERVICE

#### LUBRICATION

1. Complete chassis lubrication (see chart).

2. Lubricate accelerator linkage.

#### SERVICE

1. Change engine oil and filter.

2. Check carburator, choke and fuel filter.

3. Check timing, points (dwell) and idle speed.

4. Check engine manifold heat valve.

5. Check brake lines and brake fluid level.

6. Check oil level in differential(s).

7. Check oil level in power steering reservoir.

Date service completed:                    Signed:

**Figure 6.11**   Periodic service sheet: 250-hour.

<u>Unit No.</u>

<u>CARS</u> = <u>PICKUPS</u> = <u>JEEPS</u> = <u>BRONCOS</u>

1000 HOUR SERVICE

LUBRICATION

1. Complete chassis lubrication (see chart).

2. Lubricate accelerator linkage.

SERVICE

1. Change engine oil and filter.

2. Check carburator, choke and fuel filter.

3. Check timing, points (dwell) and idle speed.

4. Check engine manifold heat valve.

5. Check brake fluid level.

6. Change oil in differential(s).

7. Check oil level in power steering reservoir.

8. Change transmission oil.

9. Change transfer case oil (if so equipped).

10. Check brake lines, brake pads and brake linings.

11. Flush radiator. Refill with coolant and inhibitor.

12. Pack wheel bearings with grease.

Date service completed:                Signed:

**Figure 6.12**  Periodic service sheet: 1000-hour.

obviously do not apply to all the vehicles. In this case, the final sheets provide an excellent servicing routine based on hours.

There are only four sheets—daily, 50-hour, 250-hour, and 1000-hour— and these are simpler than those for the tractor. However, they were no easier to make and were probably more difficult to compose. This was because service manuals for a number of different makes and models had to be studied. Their service requirements were compared and combined so that each type was fully covered. It has been stated before and is worth repeating that it takes a lot of work and effort to arrive at a good set of periodic service sheets. It is worth the time spent, since once they are done, they almost never have to be changed.

These service sheets establish a scheduled routine with the necessary steps to provide the required service and maintenance. It does not matter much if the method of recording time worked is not always completely accurate. At times, a certain service might be off by a day one way or the other. What does matter is that there is a routine established which ensures that proper service is given at the correct intervals. A normal amount of tolerance is necessary if the system is to be practical. Few machines ever suffer because of a delay of a few hours in giving routine service. Machinery does suffer and costs skyrocket when service is neglected. This system goes a long way toward preventing neglect.

When the status boards have all been set up and all periodic service sheets are available, the routine maintenance and servicing of the equipment are fully organized. The preventive maintenance master plan is just about in place. Files have been set up with the equipment file folders. Each folder contains a maintenance history card. The master equipment inventory list has been made with the status boards and files set up in the same order and format.

The maintenance organization now knows exactly what equipment it has under its jurisdiction, what the various service requirements are for each, and it has set in place a system for recording working time for each unit. Routine periodic service can now be done on a precise schedule with strict accountability. Finally, files are in place to receive service and repair cost data.

All that remains to be done is to set up the system that will allow the service group to collect all the necessary data to compile accurate cost histories and repair information on the equipment. This is explained in detail in Chapter 7.

# chapter 7

# Completing
# the Master Plan

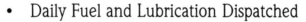

- Daily Fuel and Lubrication Dispatched
- Monthly Fuel and Lube Usage Record
- Weekly Time Distribution
- Requisition for Materials
- Repair Order
- Monthly Repair Record

The final phase of setting up the maintenance master plan is to organize the day-to-day routine of collecting repair and service data so that they can be translated into individual cost figures for each piece of equipment. There are two sources generating the data, the service group and the repair group. Each needs specialized forms.

In the preceding chapters we have described in detail how routine maintenance and servicing will be organized and managed. By its nature, the repair phase of the maintenance organization cannot be so easily defined and prearranged. Some routine scheduled work can be considered as repairs. Units in the shop for a scheduled brake system check or tire change are examples of maintenance that can be labeled as repairs. This will become evident when the different types of forms are explained. Major repairs are unplanned so are performed as the need arises. The only control that can be exercised with any degree of long-term effectiveness is a well-executed master plan for service and maintenance that will keep breakdowns and repairs to an acceptable minimum level. Ideally, the only major repair work would occur as the result of accidents or other uncontrollable circumstances.

To monitor the service and repair work done on each unit, a number of specialized forms have been devised. These have been simplified for ease in usage and to cover all types of equipment. The last thing any manager should want to do is create so much paperwork that a bureaucracy is needed to handle it. This chapter is devoted to the explanation of how each form is made up and used. Next, it will be demonstrated how each relates to the next and how the data end up on the permanent maintenance history cards. This system requires very little extra effort to make it work. The main point to remember is that all work and materials must be charged to those individual units receiving service or repairs. To be able to do this, each unit has to have a company number assigned to it that doubles as an account number against which the appropriate charges are made.

## DAILY FUEL AND LUBRICATION DISPATCHED

Figure 7.1 shows an example of the front side of this form. To withstand rough field usage, this should be printed on a grade of paper somewhat heavier than regular writing paper. Sheets should be made up in tablet form for easier handling. Tablets of this form are to be issued to each service area, lubrovan, fuel truck, shop, and person who will be providing fuel and lube service to equipment. These forms are not used by the general stores or warehouse. When oil and fuel are taken on by trucks or service stations,

# DAILY FUEL AND LUBRICATION DISPATCHED

TRUCK/AREA _____

DATE _____ 19 ___

| CO NUMBER | DIESEL | GAS-OLINE | ROTELA T MOTOR OIL 30 | ROTELA MOTOR OIL 40 | MOTOR OIL 40 | TRANSMISSION | | | | HYDRAULIC | | FINAL DRIVES | | | ANTI-FREEZE |
|---|---|---|---|---|---|---|---|---|---|---|---|---|---|---|---|
| | | | | | | AUTO TYPEA | SAE 30 | 90 | TELUS 51 | SAE 30 | DONAX T 4 | 30 | 90 | 303 | |
| | | | | | | | | | | | | | | | |
| | | | | | | | | | | | | | | | |
| | | | | | | | | | | | | | | | |
| | | | | | | | | | | | | | | | |
| | | | | | | | | | | | | | | | |
| | | | | | | | | | | | | | | | |
| | | | | | | | | | | | | | | | |
| | | | | | | | | | | | | | | | |
| | | | | | | | | | | | | | | | |

**Figure 7.1**  Daily fuel and lubrication dispatched: front.

they do not have to be charged for these products. Fuel and oil carried by service units or stations are considered part of the overall inventory. Only those pieces of equipment that use and consume these products are charged each time they are serviced. In Chapter 5 the inventory of petroleum products was discussed. All areas and equipment that store, or contain, these products are inventoried at the same time. This includes lubrovans, fuel trucks, and service stations.

The front side of the form must be filled in at the top right corner with the current date and the name of the service station or service truck number. As each unit is serviced, the company number of the unit is entered in the left column. The amount of fuel and/or lubricants are entered in the appropriate columns to the right. These are entered as gallons or fractions of gallons. For example, 0.25 would indicate a quart of oil. For those on the metric system, liters are used. In the example shown in Fig. 7.1, the company has listed all the types of products normally used and has left some blank spaces. This is to allow for the addition of different products that might be required by a new machine. Notice that 30 WT motor oil appears more than once. Some units may use this oil in other components as well as engines. Where possible, it is a good idea to separately list motor, transmission, hydraulic, final drive, and other oils to have a better picture of their use. In the case where a tractor might use 30 WT engine oil for more than one component, simply noting the total usage of this oil might give a distorted indication of the use pattern. It might give the mistaken impression that the engine oil usage rate was abnormally high when it was actually normal. Hydraulic oil changes might use a lot of the 30 WT oil.

When setting up this form, keep in mind that it is to be used to record the lube oils and fuels dispatched to all equipment in use. Therefore, it will be necessary to list all types of fuel used, the various engine oils used, different transmission, hydraulic, gear, final drive, and any other lubricants that might be used in the operation. As can be seen, this is a good argument for standardizing on lubricating oils wherever possible.

There are 26 lines on this sheet and they may not be enough for a large operation. If more sheets are needed for a day, simply number them at the top. They will all have the same date and truck number or station description. Many times, the same unit may appear more than once on a sheet. Usually, this happens when refueling is required more than once a day. The lubrovans and fuel trucks will also appear in the left-hand column, as they must also be fueled periodically. Lubrovans can also provide lube service to themselves. If hours or mileage is to be recorded, another column can be provided. Each company must make its forms to fit its particular needs, fuels, and lubricating oils.

Grease is not mentioned here because it is next to impossible to allocate it to individual units with any degree of accuracy. Grease can be considered a maintenance overhead item. It would take a great many units quite a while to use up a 400-pound drum of grease. An exception to this might be a stationary unit with an automatic greasing system. If the reservoir is a drum of grease, that can be charged to the unit when it is first set in place.

The reverse side of this form is shown in Fig. 7.2. The top half is used to keep a daily inventory of petroleum products carried by lubrovans or service stations. It is self-explanatory. Columns are provided for all products

## DAILY BALANCE OF FUEL AND LUBRICANTS

Date _____ 19 ___

|  | DIESEL | GASOLINE | ROTELA T MOTOR OIL 30 | ROTELA MOTOR OIL 30 | DONAX T4 HYDRA. OIL | MOTOR OIL 40 | AUTO. TRANS. TYPE A | 90 | TELUS 51 | 303 |  |
|---|---|---|---|---|---|---|---|---|---|---|---|
| PREVIOUS BALANCE |  |  |  |  |  |  |  |  |  |  |  |
| RECEIVED |  |  |  |  |  |  |  |  |  |  |  |
| TOTAL |  |  |  |  |  |  |  |  |  |  |  |
| TOTAL DELIVERED |  |  |  |  |  |  |  |  |  |  |  |
| BALANCE ON HAND |  |  |  |  |  |  |  |  |  |  |  |

## LUBE AND FILTER CHANGES

| CO. NUMBER | MOTOR OIL | TRANS OIL | TRANS. CASE OIL | DIFFERENTIAL | FINAL DRIVES | AIR FILTER | ENG OIL FILTER | HYD. FILTER | TRANS FILTER | FUEL FILTER |
|---|---|---|---|---|---|---|---|---|---|---|
|  |  |  |  |  |  |  |  |  |  |  |
|  |  |  |  |  |  |  |  |  |  |  |
|  |  |  |  |  |  |  |  |  |  |  |
|  |  |  |  |  |  |  |  |  |  |  |
|  |  |  |  |  |  |  |  |  |  |  |
|  |  |  |  |  |  |  |  |  |  |  |
|  |  |  |  |  |  |  |  |  |  |  |

**Figure 7.2** Daily fuel and lubrication dispatched: back.

used plus extra blank columns. Not all companies use this portion of the form, as some consider it redundant and extra work. However, it is there in case the company wants to conduct periodic checks to try to determine the source of discrepancies that might show up during the regular monthly inventory.

The bottom half is useful for indicating lube oil and filter changes. This is particularly true if the change occurs outside the normal cycle of service. For example, a case of fuel dilution may have been detected. After the fuel system was checked and the problem corrected, the crankcase oil and filter would have to be changed. This would bring it to the attention of the service supervisor and clerk.

The company number is entered in the left-hand column. Next, the appropriate box under one of the component columns is checked. It is not necessary to enter the amount of oil used since it is already entered on the front of the form. In the box(s) under the appropriate type filter enter 1 or 2, depending on how many are used. Some systems use more than one filter.

Each station, truck, lubrovan, and so on, will fill out these sheets each day, and for those units requiring oil changes, additional notations will be made on the back side. These changes may be triggered by periodic service sheets issued that day or as the result of repairs completed.

At the end of the day, or the first thing in the morning, all the *daily fuel and lubrication dispatched* sheets are to be turned in to the service supervisor. The clerk will add any multiple entries noted on the sheets so that each unit serviced that day has a single total, in gallons, for each of the amounts of fuel and lube oils received. The service clerk must also be on the lookout for units that received fuel and oils from different service areas on the same day. Those company numbers could appear on two or more sets of sheets. Once the totals are tabulated, the service supervisor, or clerk, can then move the tracking pegs on the status boards forward by an amount equal to the time worked by each unit for the previous day. This is done by dividing the fuel dispatched to the unit by the fuel rate, in gallons per hour.

Some companies may prefer to use hour meter readings or odometer readings taken directly off the equipment. If this is the case, another column must be added to the front side of the form next to the company number column. The readings can be recorded by the person giving the service. The service clerk must then subtract the previous day's reading to obtain the hours or mileage run for the day. This can be cumbersome and complicated if multiple fuelings have occurred during the day. The clerk must check to ensure that the highest reading for the day is used. The data from these daily forms are then transferred to the monthly record, described next.

## MONTHLY FUEL AND LUBE USAGE RECORD

The monthly fuel and lube usage record (Fig. 7.3) is used to compile the usage of fuel and lubricants and arrive at a monthly total for each product used. The form is printed on one side and should be in tablet form. A separate sheet is used for each unit. The unit's company number is written in the space provided at the upper left corner. Directly below that is a space to enter the month. To the right, under the title, is a line that can be used to

# MONTHLY FUEL AND LUBE USAGE RECORD
### CHANGE ENG. OIL & FILTER EACH

CO. NUMBER

MONTH _____

SERVICE BY:

| | MILES | HOURS | FUEL CONSUMED |
|---|---|---|---|
| | MILES | HOURS | FUEL CONSUMED |

| DAY | HOURS OR MILES | DIESEL GALS. | GAS-OLINE | T MOTOR OIL 30 | MOTOR OIL 30 | MOTOR OIL 40 | DONAX T-4 | AUTO TYPE A | 90 | TYPE 303 | TELUS 51 | ANTI FREEZE |
|---|---|---|---|---|---|---|---|---|---|---|---|---|
| 1 | | | | | | | | | | | | |
| 2 | | | | | | | | | | | | |
| 3 | | | | | | | | | | | | |
| 4 | | | | | | | | | | | | |
| 5 | | | | | | | | | | | | |
| 6 | | | | | | | | | | | | |
| 7 | | | | | | | | | | | | |
| 8 | | | | | | | | | | | | |
| 9 | | | | | | | | | | | | |
| 10 | | | | | | | | | | | | |
| 11 | | | | | | | | | | | | |
| 12 | | | | | | | | | | | | |
| 13 | | | | | | | | | | | | |
| 14 | | | | | | | | | | | | |
| 15 | | | | | | | | | | | | |
| 16 | | | | | | | | | | | | |
| 17 | | | | | | | | | | | | |
| 18 | | | | | | | | | | | | |
| 19 | | | | | | | | | | | | |
| 20 | | | | | | | | | | | | |
| 21 | | | | | | | | | | | | |
| 22 | | | | | | | | | | | | |
| 23 | | | | | | | | | | | | |
| 24 | | | | | | | | | | | | |
| 25 | | | | | | | | | | | | |
| 26 | | | | | | | | | | | | |
| 27 | | | | | | | | | | | | |
| 28 | | | | | | | | | | | | |
| 29 | | | | | | | | | | | | |
| 30 | | | | | | | | | | | | |
| 31 | | | | | | | | | | | | |
| TOTAL | | | | | | | | | | | | |
| PRICE/GAL. | | | | | | | | | | | | |
| TOTAL PRICE | | | | | | | | | | | | |

FUEL RATE

**Figure 7.3**   Monthly fuel and lube usage record.

indicate the normal engine oil change cycle. If this unit calls for a change every 250 hours, a "250" is written above the word "hours." This is for reference only and is not absolutely necessary. This company believed that it was a good idea to have the information noted on the monthly records as well as on the status boards, as a hedge against its being overlooked. Directly below is a line that indicates on what basis service is given. One of the three—miles (kilometers), hours, or fuel consumed—must be checked. This determines what will be entered in column 1, miles or hours, and whether column 1, 2, or 3 will be used as the basis for providing periodic service.

There are 31 lines, each representing a day of the month. Across the top are listed the various fuels and lubricants used by the company plus a column for antifreeze. These forms are designed so that they can be used for all the various types of equipment. Using a universal form eliminates the need for different monthly records for each type of equipment. The cost of printing is thus kept to a minimum by standardization, and time and space are saved by not having to deal with multiples of the same basic record. If hours are used, cross through the word "Miles." The service clerk enters the hours for that unit in the first column next to the appropriate day. Next, the clerk enters the totals from the daily fuel and lubrication dispatched sheets in the proper columns.

As with the daily sheets, it is advisable to separate lubricants by indicating where they are used. Notations should be made indicating transmission, hydraulic, final drive, and so on, over the oil columns. Motor oil is already noted in this example. Referring back to Chapter 6 and the *periodic service sheets* of the typical heavy tractor, we find that the types of lube oils used are noted on these sheets. This information can also be found on the data panel of the tractor's *equipment file folder*. This typical tractor uses the following lubricants:

| | |
|---|---|
| Motor oil | SAE 30 |
| Hydraulic oil | SAE 30 (motor oil) |
| Transmission | ATF-A |
| Transfer case | EP 90 gear oil |
| Final drives | SAE 30 (motor oil) |

In the sample monthly record shown in Fig. 7.3, the motor oil dispatched might be listed in the column titled "T Motor Oil 30." The next column can be used for hydraulic oil. With a pen, cross out the word "Motor" and write the word "Hydraulic" over it. Transmission oil would go in the column already titled "Auto Type A" and transfer case oil in the column titled 90. For final drives, use any free column by crossing out the title and relabeling it "Final Drive 30." Figure 7.4 shows what the sheet would look like with a typical entry for the first day of the month.

It matters little if the sheet appears marked up in some of the columns. Another type of unit might use other oils listed on the form. These forms are not a permanent record. They are kept in the equipment file folder during the month. At month's end, the data are transferred to the *maintenance history card* and the monthly form can be thrown away. A new form

# MONTHLY FUEL AND LUBE USAGE RECORD

### CHANGE ENG. OIL & FILTER EACH

_502_
CO. NUMBER          _____ MILES          _250_ HOURS          FUEL CONSUMED

SERVICE BY:

MONTH _Jan. 1986_          _____ MILES          ✓ HOURS          FUEL CONSUMED

| DAY | HOURS ~OR~ ~MILES~ | DIESEL GALS. | GAS-OLINE | T MOTOR OIL 30 | ~Hydraulic~ ~MOTOR~ OIL 30 | MOTOR OIL 40 | DONAX T-4 | ~Trans.~ AUTO TYPE A | Trans Case 90 | Final Drive ~TYPE~ ~303~ 30 | TELUS 51 | ANTI FREEZE |
|---|---|---|---|---|---|---|---|---|---|---|---|---|
| 1 | 16 | 100 | | .25 | — | | | — | — | — | | |
| 2 | | | | | | | | | | | | |
| 3 | | | | | | | | | | | | |
| 4 | | | | | | | | | | | | |
| 5 | | | | | | | | | | | | |
| 6 | | | | | | | | | | | | |
| 7 | | | | | | | | | | | | |
| 8 | | | | | | | | | | | | |
| 9 | | | | | | | | | | | | |
| 10 | | | | | | | | | | | | |
| 11 | | | | | | | | | | | | |
| 12 | | | | | | | | | | | | |
| 13 | | | | | | | | | | | | |
| 14 | | | | | | | | | | | | |
| 15 | | | | | | | | | | | | |
| 16 | | | | | | | | | | | | |
| 17 | | | | | | | | | | | | |
| 18 | | | | | | | | | | | | |
| 19 | | | | | | | | | | | | |
| 20 | | | | | | | | | | | | |
| 21 | | | | | | | | | | | | |
| 22 | | | | | | | | | | | | |
| 23 | | | | | | | | | | | | |
| 24 | | | | | | | | | | | | |
| 25 | | | | | | | | | | | | |
| 26 | | | | | | | | | | | | |
| 27 | | | | | | | | | | | | |
| 28 | | | | | | | | | | | | |
| 29 | | | | | | | | | | | | |
| 30 | | | | | | | | | | | | |
| 31 | | | | | | | | | | | | |
| TOTAL | | | | | | | | | | | | |
| PRICE/GAL. | | | | | | | | | | | | |
| TOTAL PRICE | | | | | | | | | | | | |

FUEL RATE [                    ]

**Figure 7.4**   Monthly fuel and lube usage record (filled out).

for the next month is immediately started for the unit, and it will have the same marked-up column titles.

When listing hours or miles, it is for the service group to decide whether or not to use cumulative totals. Since cumulative totals are used on the status boards, the author prefers to use the actual time or distance logged each day. At the end of the month these can be added to give the total hours worked, or miles traveled, during the month. If running totals were used, the first entry for the next month might be 400 hours or perhaps 5210 miles, which is meaningless for a single entry on this particular record.

As soon as the monthly records are completed through the final day, add each column and enter the totals in the spaces provided. Usually, these totals will be in gallons. Next, determine the unit price paid for each of these products. The purchasing department can provide this information. Enter the prices in the appropriate spaces. Now calculate the total price (cost) of the fuel and each lubricating oil used during the month. If antifreeze and corrosion inhibitor are used, they should be totaled and priced in the same way.

By dividing the total fuel used during the month by the total hours worked, the fuel rate is established for the month for that particular unit. If the fuel rates for similar units are averaged, the service supervisor can compare this to the rate initially established for calculating working hours used on the status board. If warranted, the supervisor can adjust the rate used on the boards. With time and experience, he or she will be able to determine what is a good practical rate, as opposed to the theoretical rate, and fine tune the system accordingly.

The monthly cost totals are transferred to the maintenance history card and listed, for the appropriate month, in the section titled "Cost of Fuel and Lubricants." For this typical tractor, the costs of transmission and transfer case oils are added together, as they are both considered to be transmissions. If there is a cost for antifreeze and rust inhibitor, enter it in the fuel column not used. In this case, the typical tractor is diesel powered, so the word "Gasoline" should be crossed out and "Antifreeze" written in. Add the costs and enter the total in the space marked "Total Cost Fuel & Lub." Refer to Fig. 4.3, which illustrates the maintenance history card.

On the reverse side of the maintenance history card, enter the total hours, or mileage, in the first column. This should show the cumulative totals as opposed to the monthly fuel and lube usage record, which lists only day-to-day hours worked or mileage run. For the second month, add the total to that of the first month. The maintenance history card is a permanent record, so the cards should show the total hours worked, or mileaged logged, for the life of the unit. For the next year, the first month's entry will include the hours from month 12 of the previous year. When starting this system with existing units, try to determine the total hours worked, or mileage, and add this amount to that of the first reporting month.

Once all this information has been entered on the maintenance history card, the monthly fuel and lube usage record can be destroyed. While in use, keep this form in the equipment file folder.

This covers the steps for collecting cost data on fuel and lube service given to the equipment. The cost of filters has not been discussed. How this is recorded is explained later in this chapter.

In explaining the makeup and use of forms, we have stressed the importance of keeping things as simple as possible and still be flexible enough to make allowances for the different needs of the equipment. A company that has only a few different types of equipment will find it much easier to design forms than one with many varied types of operating units in its inventory. In both cases the first thing to do is make a list of all the fuels and lubricants used. Next, group the lubricants by service, such as motor oils, transmission oils, hydraulic oils, and so on. At this point it is worth while to investigate the possibility of standardizing on fewer types of lubricants. Once a standard list of fuel and lubricants is agreed to, the forms can be designed around them. Do not have forms printed with specific name brands. In the future, the company may change suppliers and the forms will become obsolete. Use generic names or generally understood descriptions.

The other type of data required concerns repair costs and other shop-related work done on the equipment. The foremen and supervisors of the repair group are responsible for providing this information to the service group. To do this, they need forms designed for their particular needs. The cost of labor and materials must be known in order to establish the shop cost of repair work.

## WEEKLY TIME DISTRIBUTION

This form is used to allocate the time spent by hourly workers in the repair group. As a general rule, the service technicians servicing machinery do not record work time spent on each unit serviced. Instead, their time is considered part of overhead. This is because they spend such a relatively short time on any one piece of equipment that it is not worth the effort trying to assign fractions of labor hours to units receiving lube service. Forty to fifty percent of their time may also be taken up moving from unit to unit if mobile equipment is involved.

The *weekly time distribution* form is used in addition to the clock printed time card used for payroll purposes. Weekly totals from the time card are used by the payroll department to compute the pay of the employees for the week. The fact that an hourly employee may have performed work on a number of different units is not taken into consideration. The total cost of wages is charged to the maintenance department, but what the department needs to know is exactly how this time was spent during the week. One simple way to keep track of the time spent on each repair job is to use a time distribution sheet. Figure 7.5 is an example of how one company chose to record repair times, and it very simply illustrates the way to do it.

These forms can be printed in tablet form and are to be filled out by the foremen or supervisors. A separate sheet is used for each hourly employee. The person's name and employee number are listed at the top, and the week ending date is also noted.

The left-hand column is for listing the units worked on by entering the appropriate company (unit) numbers. Remember that a unit's company number doubles as an account number. There may be times when an employee will be working at something other than repairs to a particular unit.

## WEEKLY TIME DISTRIBUTION

Emp. Name _____    No. _____

Week Ending _____

| Co. No. | Mon. | Tue. | Wed. | Thu. | Fri. | Sat. | Sun. | Tot. | Comments |
|---------|------|------|------|------|------|------|------|------|----------|
|         |      |      |      |      |      |      |      |      |          |
|         |      |      |      |      |      |      |      |      |          |
|         |      |      |      |      |      |      |      |      |          |
|         |      |      |      |      |      |      |      |      |          |
|         |      |      |      |      |      |      |      |      |          |
|         |      |      |      |      |      |      |      |      |          |
|         |      |      |      |      |      |      |      |      |          |
|         |      |      |      |      |      |      |      |      |          |
|         |      |      |      |      |      |      |      |      |          |
|         |      |      |      |      |      |      |      |      |          |
|         |      |      |      |      |      |      |      |      |          |
|         |      |      |      |      |      |      |      |      |          |
|         |      |      |      |      |      |      |      |      |          |
|         |      |      |      |      |      |      |      |      |          |
|         |      |      |      |      |      |      |      |      |          |
|         |      |      |      |      |      |      |      |      |          |
|         |      |      |      |      |      |      |      |      |          |
| Total   |      |      |      |      |      |      |      |      |          |

_____    _____
Authorized By                    Date

**Figure 7.5** Weekly time distribution.

An electrician might spend the entire week working on the electrical system of a shop building. This should also be covered by an account number for physical plant upkeep and repair. So this would be the number that is listed in the first column.

Each day, enter the hours spent on each job or unit. The total at the bottom will show the total time worked for the day. If the employee should work overtime, record the excess time as straight time. For example, if a mechanic works 1 hour of overtime on Monday and gets time-and-a-half for that hour, record the overtime as 1.5 hours. The total for Monday will be 9.5 hours. This makes it easier when applying pay rates to arrive at labor costs on a particular repair job. The total column at the right contains the total hours worked on each job for that week.

The area under the heading comments can be used to note a brief description of the nature of the work and to note the person's pay rate. At week's end, the supervisor signs and dates the sheets and turns them in to the repair group clerk.

The example shown in Fig. 7.5 is very simple but contains all the information needed to charge labor costs to individual units. Maintenance organizations with differing amounts of equipment to care for and a differing work force may need to use a modified version of this form. Some industries use the latest types of electronic time clocks. Some of these digital units, with keys for recording job numbers, or account numbers, are interfaced with a computer. The employee enters the number of the job being worked on and uses his or her time card to punch in. When the job is finished, the employee immediately punches out. He or she then enters a new job number and punches in again. The computer keeps track of the total time for payroll purposes and automatically charges labor costs to the various repair jobs.

A wide range of systems are available for assigning accurate labor costs to repair jobs and unit account numbers. It is recommended that the simplest, most economical method that is consistent with the organization and its capabilities be used.

## REQUISITION FOR MATERIALS

The existing form used by a company can usually be retained with little change in the overall format. Figure 7.6 illustrates a requisition that is commonly used. There must be spaces to list the account number, or unit number, of the machine and the cost of the part. The other columns and spaces are self-explanatory.

These should be made up in tablet form with carbons so that a copy can be retained by the repair or service group. The completed form is taken to the warehouse, or general stores, and the order filled by the warehouse personnel, who will list the prices on the form. The repair group will retain the copy and turn it in to the repair clerk. Parts listed on the requisition should be for one account number only. Use a separate form for each different unit or account number.

If the company uses a computer with terminals in each department and has the stock listing with costs available for reference, the repair group clerk can use this information to fill in part costs when making up the requisition. Regardless of how it is done, it is essential that the price information and correct account number are entered so that accurate charges can be allocated.

## REPAIR ORDER

Refer to Fig. 7.7 for an example of a repair order. These forms should be made out whenever repairs are to be made, either in the shop or on location. The unit data must be entered at the top and signed by an authorized

| REQUISITION FOR MATERIALS | | | | | |
|---|---|---|---|---|---|
| CO. NO. _____ (Acct. No.) | | | DATE _____ | | |
| Qty. | Code | Part No. and Description | unit cost | Total |
| | | | | |
| | | | | |
| | | | | |
| | | | | |
| | | | | |
| | | | | |
| | | | | |
| | | | | |
| | | | | |
| Authorized By | | | Dispatched By | | |

**Figure 7.6** Requisition for materials.

supervisor. The company number of the unit is the account number against which the cost of parts and labor are charged. Next, an explanation of the repairs to be performed is written in the space provided. If the order is to cover diagnostic work to determine the nature of a malfunction, this should be noted along with the listing of symptoms. Once the problem has been isolated, the repair procedure can be outlined briefly in the same space.

These sheets should be made up in tablet form on standard letter-sized paper. A copy should be given to the repair group clerk as soon as the explanation of repairs section has been filled in with the pertinent information. The supervisor should keep the original and provide a second copy to the person who will actually perform the work on that unit.

As the work progresses, the repair personnel will order parts as required, noting them on the right side of the form. Copies of these completed requisitions, together with copies of the weekly time distribution sheets, will be turned in to the repair group clerk. This person will be responsible for posting all material and labor costs on his or her copy of the repair order. When a repair job requires a lot of labor and materials, the reverse side of the repair order form can be used to list these charges. The totals will be entered on the front side when the job is completed.

The job can be closed out when the clerk receives the supervisor's copy of the repair order, signed and dated as completed. The clerk will check with the supervisor to ensure that all the materials and labor expended on that particular job are listed on the repair order. Once the repair order is

# REPAIR ORDER

Make _____ Model _____ Co. Number _____

Serial Number _____                     Date _____

Hours or Miles _____                    Authorized Signature _____

| ITEM | EXPLANATION OF REPAIRS | MATERIALS | QUANTITY |
|---|---|---|---|
| | | | |

| TOTAL MATERIALS COST | LABOR — HOURS COST |
|---|---|
| | RATE |
| | _____ _____ |
| | _____ _____ |
| | _____ _____ |
| | TOTAL _____ |

_____
DATE FINISHED

_____
SIGNED

**Figure 7.7**  Repair order.

closed out, it is given to the service clerk for entry into the cost record for that unit.

There may be instances when repairs are done by outside contractors or dealers. When this occurs, use the invoice as the source of cost information and repair details. Give a copy of the invoice to the service clerk so that these data can be entered in the permanent record of the unit. Most invoices will list labor and materials separately, so this presents no problem. If the charges are not broken down, contact the firm that performed the work and request an itemized invoice. In cases where a dealer performs warranty repairs, there will not be any cost involved, but a description of the work done must be entered into the unit's record.

## MONTHLY REPAIR RECORD

The monthly repair report is used to compile the repair cost data for the month, individually for each machine, and is used much like the monthly fuel and lube usage report. It is usually simpler in format because there are fewer diverse data to record during any given month. Figure 7.8 shows an example of this form. The *monthly repair record* is normally printed on letter-sized paper, as are a majority of the forms described in this book. Making them available in tablet form eases handling and storage problems. One form is used for each unit, with its company number and the month entered at the top.

This sheet is kept in the unit's equipment file folder together with the maintenance history card and the monthly fuel and lube usage record. However, instead of automatically starting a new sheet at the beginning of each month for every unit, a monthly repair record is started only when repairs or consumable service parts are charged to the unit.

The left-hand column is for entering the date the repair work was finished. If a repair job is begun in one month but continues into the following one, the date entered on the record is the day of the second month when the repairs were actually completed. For the sake of simplicity, no data for the job are entered during the month in which work was started.

In the next space, a brief description of the repairs is noted. This is preferably a one-line explanation. The total cost of materials and labor for that job is entered in the last two columns. This information is taken from the completed *repair order* submitted by the repair group clerk.

There are other types of entries in addition to those resulting from completed repair orders. There may be work done by outside contractors and dealers. The information is entered in the same way as for repairs done in-house. Warranty repairs should also be recorded. List the date completed and a description of the failure and repairs required. There may be no charge, in which case the word "none" or "warranty" is written through the last two columns. Do ensure that all warranty work is labeled as such and that a copy of the invoice, or warranty report, is kept in the equipment file folder in addition to the other records.

There is one other type of entry made on these monthly forms and that is the cost of the odd part, usually a consumable, purchased for the unit. An example is the oil filter used when providing lube service to the equipment.

# MONTHLY REPAIR RECORD

_____                    _____
CO. NUMBER                                              MONTH

| DATE WORK FINISHED | BRIEF EXPLANATION OF WORK DONE | SHOP COSTS | |
| --- | --- | --- | --- |
| | | MATERIALS | LABOR |
| | | | |
| | | | |
| | | | |
| | | | |
| | | | |
| | | | |
| | | | |
| | | | |
| | | | |
| | | | |
| | | | |
| | | | |
| | | | |
| | | | |
| | | | |
| | | | |
| | | | |
| | | | |
| | | | |
| | | | |
| | | | |
| | | | |
| | | | |
| | | | |
| | | | |
| | | | |
| | | | |
| | | | |
| | | | |
| | | | |
| | | | |
| | | | |
| | | | |
| | | | |
| | | | |
| | | | |
| | | | |
| | | TOTAL | |

Figure 7.8   Monthly repair record.

When a filter is purchased by service personnel for use in routine maintenance, a *requisition for materials* is completed. As soon as the filter is in hand, the copy of the requisition is given to the service clerk, who enters the information on the unit's monthly repair record. This consists of the date, part name, and the cost of the filter. Labor, in this case, is considered as department overhead and so is not entered as a direct cost.

At month's end, the totals of material and labor costs are entered at the bottom of each column. There are sufficient lines on this record sheet for listing repairs and parts for the month. There may be a high percentage of equipment that has nothing to report for a given month—thus the reason for not starting a monthly repair record on a unit until something is reported. This is as it should be. The maintenance department should have as its goal the reduction of repair jobs to an absolute minimum.

The monthly cost totals from this record are transferred to the maintenance history card and listed for the appropriate month. These totals are entered in the columns in the section titled "Repair Costs." Add the material cost to the labor cost and enter this in the column for "Total Cost Repairs." This total, combined with the "Total Cost of Fuel and Lubricants," is the total cost for the month shown in the last column. Refer to Fig. 4.3 for an illustration of the maintenance history card.

Some maintenance departments may prefer to apply a departmental overhead figure to these totals. If this procedure is adopted, add a note below the word "Total" in the last column indicating that the total includes a designated percentage for overhead. Whether shop cost or full departmental cost is used is up to management to decide. Make sure that whichever is used, it is used throughout the system. Do not have a mix of two cost methods, or it will be impossible to obtain reliable comparative cost studies.

On the reverse side of the maintenance history card, enter any pertinent information concerning repairs for the appropriate month. Do not bother entering minor items such as filter changes or battery replacement. Do list work, under the appropriate headings, that would be considered bona fide repairs. Included in this category would be all warranty work, even though it might not involve charges. The purpose of the maintenance history card is to record the cost and repair histories of individual units. Over a period of months, or years, patterns of breakdowns and repair problems may become evident. These may show up in single units or may indicate a common trend in a number of units of the same make and model. Similarly, the cost data, compared with hours worked, may also show trends in the equipment. All of these data are extremely valuable to the maintenance manager as well as the general management. This historic data base will enable management to arrive at sound budgeting techniques and will also form the basis for decisions concerning the purchase of new capital equipment.

Figure 7.9 illustrates what a completed monthly repair record might look like with costs totaled. Once all the information is transferred to the maintenance history card, this record can be destroyed. To illustrate the system further, Figures 7.10 and 7.11 show the front and reverse sides of a typical maintenance history card completed for the first month.

The procedures described in this chapter complete the system referred to as the *preventive maintenance master plan*. All the forms and procedures have two purposes: to organize the service and repair of equipment and to

# MONTHLY REPAIR RECORD

502
CO. NUMBER

Jan. 1986
MONTH

| DATE WORK FINISHED | BRIEF EXPLANATION OF WORK DONE | SHOP COSTS | |
| --- | --- | --- | --- |
| | | MATERIALS | LABOR |
| 1-15-86 | Repair Radiator leak | 5.89 | 39.50 |
| 1-28-86 | Two Oil Filters for engine oil change | 22.80 | |
| 1-30-86 | Replace faulty injector (Warranty) | — None — | |
| | | | |
| | | TOTAL | $28.69 | $39.50 |

**Figure 7.9** Monthly repair record (filled out).

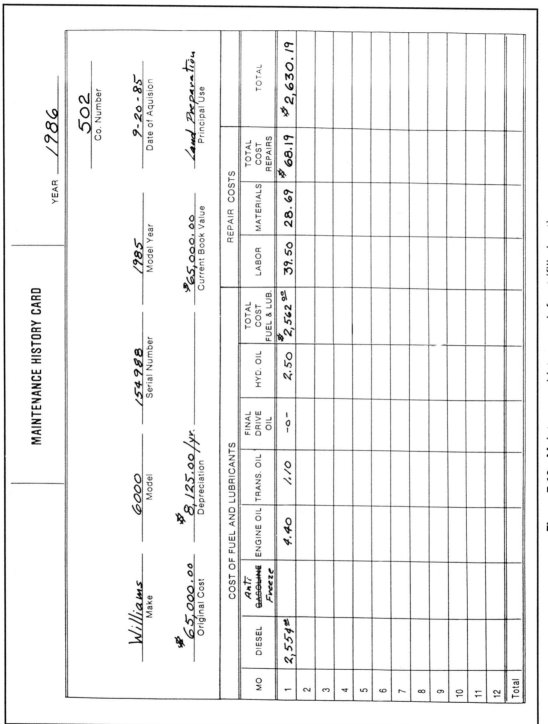

**Figure 7.10** Maintenance history card: front (filled out).

## EXPLANATION OF REPAIRS

YEAR  1986

| MO. | HOURS/ MILES | ENGINE | TRANSMISSION | FINAL DRIVES | ELECTRICAL SYSTEM | HYDRAULIC SYSTEM | OTHER |
|-----|-----|-----|-----|-----|-----|-----|-----|
| 1 | 688 | Fuel Injector Replacement (warranty) | | | | | Radiator Repaired |
| 2 | | | | | | | |
| 3 | | | | | | | |
| 4 | | | | | | | |
| 5 | | | | | | | |
| 6 | | | | | | | |
| 7 | | | | | | | |
| 8 | | | | | | | |
| 9 | | | | | | | |
| 10 | | | | | | | |
| 11 | | | | | | | |
| 12 | | | | | | | |

**Figure 7.10**  Continued (back of form).

EXPLANATION OF REPAIRS

YEAR  1986

| MO. | HOURS/MILES | ENGINE | TRANSMISSION | FINAL DRIVES | ELECTRICAL SYSTEM | HYDRAULIC SYSTEM | OTHER |
|---|---|---|---|---|---|---|---|
| 1 | 688 | Fuel Injector Replacement (Warranty) | | | | | Radiator Repaired – Accident caused leak |
| 2 | | | | | | | |
| 3 | | | | | | | |
| 4 | | | | | | | |
| 5 | | | | | | | |
| 6 | | | | | | | |
| 7 | | | | | | | |
| 8 | | | | | | | |
| 9 | | | | | | | |
| 10 | | | | | | | |
| 11 | | | | | | | |
| 12 | | | | | | | |

**Figure 7.11**  Maintenance history card: back (filled out).

# MAINTENANCE HISTORY CARD

YEAR _____

Make _____
Model _____
Serial Number _____
Model Year _____
Co. Number _____
Date of Aquision _____

Original Cost _____
Depreciation _____
Current Book Value _____
Principal Use _____

## COST OF FUEL AND LUBRICANTS

| MO | DIESEL | GASOLINE | ENGINE OIL | TRANS. OIL | FINAL DRIVE OIL | HYD. OIL | TOTAL COST FUEL & LUB. |
|----|--------|----------|------------|------------|-----------------|----------|------------------------|
| 1  |        |          |            |            |                 |          |                        |
| 2  |        |          |            |            |                 |          |                        |
| 3  |        |          |            |            |                 |          |                        |
| 4  |        |          |            |            |                 |          |                        |
| 5  |        |          |            |            |                 |          |                        |
| 6  |        |          |            |            |                 |          |                        |
| 7  |        |          |            |            |                 |          |                        |
| 8  |        |          |            |            |                 |          |                        |
| 9  |        |          |            |            |                 |          |                        |
| 10 |        |          |            |            |                 |          |                        |
| 11 |        |          |            |            |                 |          |                        |
| 12 |        |          |            |            |                 |          |                        |
| Total |     |          |            |            |                 |          |                        |

## REPAIR COSTS

| LABOR | MATERIALS | TOTAL COST REPAIRS | TOTAL |
|-------|-----------|--------------------|-------|
|       |           |                    |       |

**Figure 7.11** Continued (front of form).

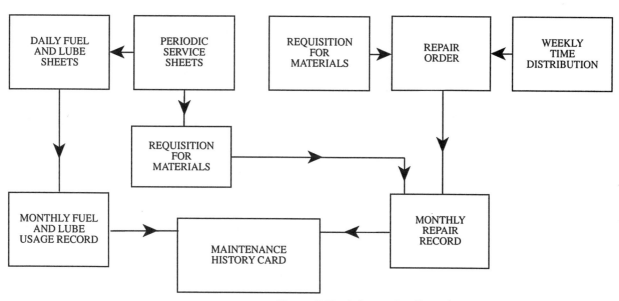

**Figure 7.12**   Information flow chart.

collect accurate cost data for each unit. Figure 7.12 summarizes the use of these forms and how they relate to each other by the use of a simple flowchart.

The forms and records discussed in this chapter can be destroyed as soon as the information they contain is transferred to the maintenance history card. This card is the only permanent record of individual units and is kept in the equipment file folder. These cards, each containing the data for one year, are kept until the unit is taken out of service permanently. If the unit is sold or traded in on a new piece of equipment, the file should go with it.

During the month, the monthly fuel and lube usage record and the monthly repair record are kept in the equipment file folder and maintained by the service personnel. Also, completed repair orders may be stored in these file folders until the service clerk transfers the data to the monthly record. The best policy is to transfer all data as they come in and not let things pile up. However, the realities of the workplace often differ from the ideal. Some material costs are not always readily available and reports are not submitted on time. This requires some flexibility in handling the flow of information. The equipment file folder is the place for holding all information on a unit until those data can be entered in the proper records.

The repair group may wish to retain copies of completed repair orders, weekly time distributions, and material requisitions. How long these documents are kept depends on departmental policy. Once the information has been transferred to a permanent record, it is recommended that these copies not be retained for more than six months. If possible, they should be disposed of sooner. If allowed to accumulate, these documents just take up valuable office space. Experience shows that once filed, this information is rarely referred to again.

The same advice applies to the service group. Once a *periodic service sheet* has been returned signed and dated, indicating that work has been completed and the fact noted on the status boards, it can be destroyed. The service supervisor should not allow completed daily fuel and lubrication sheets to pile up. Make sure that the clerk gets this information transferred to the monthly fuel and lube usage record each day. Requisitions issued from the service group should have their data entered on the monthly repair record as soon as the item is obtained.

There is nothing more detrimental to good record keeping than having reports and information piling up unattended. At best, it causes unnecessary surges in the work load. At worst, it can result in a situation completely out of control, where information is backlogged for months. In this case, data usually get lost or recorded inaccurately.

The object of the preventive maintenance master plan is to establish organization and control over service and maintenance, so care must be taken to keep the work and information flowing smoothly. Supervisors in the service and repair groups must constantly monitor the system. They must also work closely together to ensure the accuracy and completeness of the data being passed through the system.

Throughout this chapter, the terms *account number, unit number,* and *company number* have been used. These are interchangeable and are taken to mean the same thing.

The forms and reports described in this work are normally printed on standard letter-sized paper. There are some exceptions to this, notably the requisiton for materials and the weekly time distribution. In the interest of paper and cost savings, these can be smaller, as are the monthly inventory of fuel and lubricants and the monthly report of fuel and lubricants used.

The equipment file folder is a standard legal-sized manila folder printed on one side. This folder normally contains the maintenance history card and the two monthly working records. This file folder and its use are discussed in more detail in Chapter 8.

Individual organizations probably need to modify some of the forms described to suit their particular situations. Each form must, however, contain all the basic information as outlined in the examples given throughout this book. The other point to remember is that the finished forms must be simple as well as informative. This is not as easy as it sounds. Almost anyone can create something cumbersome and complicated, but it takes work and creativity to arrive at a simple workable system.

# chapter 8

# Management and Planning

- General Observations
- Practical Use of the System
- Cost Analyses
- Budget Projections
- Special Studies
- Computer Applications

Once the preventive maintenance master plan is set up and put into routine daily use, the maintenance department manager and general management will begin to appreciate the enormous benefits that can be gained from this system. This relatively simple system has evolved from the need to organize and control the highly technical, and often confusing, function of equipment management.

To plan and move ahead, the company must first know its current status exactly. When dealing with machinery and equipment that are the company's tools of trade, this information is not as easily obtained. In many companies, the cost of owning and operating equipment is the major factor in determining the profitability of the operation. It is difficult for high-level managers and administrators to focus on the details of equipment management. In the first place, it is a field covering diverse technologies, and in the second place, there has historically been no academic avenue of instruction and learning in this field. The decision makers, individually or as a group, are knowledgeable in the areas of production, sales, marketing, finance, and law. All these business activities can be learned at the university level. Later, the skills and expertise associated with a particular profession can be expanded in the workplace. This is not so with maintenance management. Those in charge have either gained their experience over long periods and have advanced up through the ranks or are professionals who hold a degree in engineering or an applicable science. Even with a degree, it often takes years before a person can be considered a top-rated maintenance manager.

The administration must look to the maintenance manager and staff for guidance and technical input, and unless they can talk to higher management in a language understandable to all, they will be treated with some degree of suspicion. The language that is universally understood in any business is money, or profit and loss. When presenting proposals to top management for the acquisition of new equipment or disposal of old machinery, the maintenance manager may have the best technical reasons to substantiate a position. However, if that is all the person has to back up a plan, the best he or she can hope for is a sympathetic hearing.

What will generally happen after such a meeting is that the general managers will make the decision as to the best course of action. The rationale here is that they, at the top level, have a better understanding of the business and must therefore make the final decision based on their knowledge and instincts. In this scenario, they may have a 50 percent chance of being right. It is the maintenance manager who has dropped the ball and not done his job as a member of the management team. He should be able to back up his proposals with accurate cost analyses based on historical data. Management may not completely understand the technical details behind a proposal, but they all understand a balance sheet. If a course of action will ultimately produce a profit or readily pay for itself, it will be accepted. Of course, management must be certain that the cost figures are based on accurate data and must have confidence in the maintenance system.

The *preventive maintenance master plan* provides the basic structure on which to build sound cost analyses, equipment usage studies, and budget projections. The master plan, set forth in this book, has established the system that will allow the department to function efficiently, be cost effective, and provide accurate data on which to base good business decisions. In summation, the following subjects have been discussed in detail:

Organizational requirements
Personnel requirements
Surveying equipment and establishing files
Petroleum products accountability
Establishing methods of routine service and maintenance
Establishing the system for data collection

These general subject areas are the basis of a good maintenance system. There are as many possible modifications to the master plan, set forth here, as there are different companies. If an organization builds on this basic system and does not change the principles involved, it will be able to fulfill the two functions necessary for a well-run department. These functions are organization and control.

There are some observations that are appropriate to make that might not have been stressed in the preceding chapters. The first deals with lubricating oils. Smaller companies may purchase oil in drums. Since more than one or two types of oil are generally used, a problem sometimes develops with identification of the different drums. It does not take much imagination to realize what could happen if hydraulic oil were mistakenly substituted for 40 WT engine oil and put in an engine crankcase. Unfortunately, oil drums are usually labeled on the drum head in stenciled letters and numbers. These are sometimes difficult to read. To prevent identification problems, it is good practice to color code the drum heads with weather-resistant paint. Make up color indexes for the various lube oils and greases used and post them at each place where the products are stored. Use a 3- or 4-inch brush and paint a stripe across the drum head and sides. Color coding may help to avoid future repair costs and lost production time by reducing the chance of using the wrong product. The amount of color coding can be reduced by standardizing on fewer lubricants.

Another point to stress is the importance of the unit numbers assigned to machinery. These are the account numbers against which costs are charged. One of the first steps to take when starting out is to discuss with the accounting staff the method of assigning numbers to equipment. It would be an unfortunate mistake to assign unit numbers and then be informed by accounting that they do not fit the system and therefore cannot be used. Plan ahead so as not to lose time redoing the system. There will be some system already in use covering the maintenance department. Together, devise a system of account numbers based on what is in current use so that the changes introduced will have a minimal impact on the overall accounting system and the maintenance department. An example of a system used by one company is shown below.

| | |
|---|---|
| General Account Number Maintenance Department | 10900 |
| Buildings and Structures | 10910 |
| Tools and General Equipment | 10920 |
| Office Supplies | 10930 |
| Indirect Labor | 10940 |
| Miscellaneous | 10950 |

Since the 10,000 series numbers were reserved for the maintenance department, the numbers 10,100 through 10,899 could be used for numbering equipment. Only the last three digits were painted on the individual units and used throughout the maintenance system. This posed no problem for the accounting group, since it was the only department that would be using a three-digit system. Accounting would automatically assign these costs to the maintenance department, but broken down by individual units. Any other maintenance department costs would be charged to the five-digit numbers shown above. There were actually more of these than shown here. Try to limit the numbers used on the individual machines to a maximum of four digits. The larger the number, the greater the possibility of error in recording machine numbers during servicing. The three-digit numbers are best suited to this system. They are simpler to read and are very adaptable, especially where there are many different unit types to be listed. Refer to the equipment inventory list in Fig. 4.1.

The labor cost for the service technicians is normally considered as indirect labor and thus part of the department overhead. These people are the ones giving routine lube service to the various working units. Normally, they will not spend over 20 minutes on any one machine, so it is not worth while attempting to assign such small time increments to individual units. It would be a nightmare for the time clerk trying to add up and assign time by fractions of hours. Finally, times reported would probably be the result of inaccurate guesswork. No one is going to time his or her work on each machine. Add to this the fact that movement time between units may account for at least 25 percent of the day's total working time. Often, this percentage is higher. Service time can, in some cases, be charged to the unit. This might occur when a technician is involved in a procedure that takes an hour or more. In general, even this is not worth bothering with, as this evolution may take place only at widely spaced intervals of time.

Another point concerning the system should be mentioned. This deals with the use of computers. This book has described in detail the procedures for setting up and implementing a smoothly functioning maintenance system. It will be obvious to many readers that computers have a great potential application in this system. Indeed they do, but do not begin installing a maintenance system master plan by employing a computer at the outset. There will be too much to do and learn without adding mastery of computer usage and programming to the work load.

It takes time to get a maintenance system working smoothly. It takes time to get the people trained to the point of being proficient in their work. During this initial phase, there will be modifications and bugs to be worked out of the new system. Once a routine is established, the leadership of the maintenance organization will have time to step back and make an overall assessment of how the operation is functioning. This can take a year or more before they are satisfied that things are working to everyone's satisfaction and no more fine tuning is needed. The system must be initially undertaken manually to understand it fully.

It is useless to computerize a system unless that system is first thoroughly mastered down to the last detail, prior to the start of computer programming. Computers cannot be effectively programmed if the basic parameters have not been well thought out and established in advance. Later, when the system is operating smoothly, the application of a computer can be investigated. The preventive maintenance master plan set forth here could use a computer as a timesaving data accumulator. It would store the data for the monthly records and produce the maintenance history cards. Where a computer will be most useful will be for special cost studies, machinery evaluations, and budget projections. These applications are made possible due to the fact that there is now a master plan that produces accurate historical data on the equipment. Thus the greatest benefits of computer usage lie outside the limits of the basic system as described thus far. The data base accumulated by this system is the foundation upon which more sophisticated computer programs can be designed. When combined with data from other departments, studies can be continuously made concerning the efficiency and cost-effectiveness of operations and equipment.

If operator's wages are added to the data along with depreciation, it is easy to obtain the total cost of owning and operating the equipment. Individual units can be compared or types of units can be evaluated as groups. If annual hours worked are compared with the hours available per year for work, we have another indicator of how efficient the usage has been. If the total amount of repair downtime is recorded, we have another indicator. By subtracting working and downtime from available time, we have standby or dead time. This can be a major factor in determining if machinery is being used to the fullest advantage. If the types of repairs are coded for use in a computer, management can use this information to identify chronic problem areas within the machinery or the system.

All these data can be assembled for comparison in an almost infinite number of ways for studying management options concerning machinery ownership and operation. Input from other departments not only helps the maintenance manager to develop the various analyses but also ties the department into the functions of the others. For example, cost figures on machinery and equipment can be compared to tons of ore hauled, miles traveled, passenger miles, acres prepared, or kilowatts generated, to name just a few examples.

The finance department may look for ways to reduce the cost of insurance and the cost of obtaining capital equipment. The production department may strive to reduce the cost of labor or cost per unit of production. Each department tries to reduce the costs involved in producing goods or services, and the department in charge of maintaining equipment is no different. If the availability of equipment is increased with an accompanying reduction in maintenance and operating costs, other departments benefit, as does the whole company.

The maintenance department staff, with all these data available, can now take the time to stand back and evaluate the operation. By selecting some, or all, of the data and setting up spreadsheets, vital comparisons can be made and problem areas identified quickly. Once isolated, problem areas can be targeted for resolution and elimination. High rates of fuel use or engine repairs might plague certain units. Referral to records might indicate a high level of repairs caused by poor operating procedures. Data might indicate an unacceptable amount of standby time. Having the data available allows management to determine what the problem is and exactly where it is (i.e., with machinery or with personnel). Knowing exactly what the problem is usually indicates that the solution is the next, and easiest, step. Too many maintenance organizations address the symptom but not the cause and so are faced with the same recurring problem.

These data are also valuable for budgetary use, for both operating budgets and capital equipment budgets. The one item that has historically been a thorn in the side of maintenance staffs is how much money to budget for repair and maintenance each year. Now that a maintenance master plan is in place and working, this problem has all but been eliminated. Even after only one year, the exact costs are known. As time goes on and management control takes full effect, the costs will actually begin to decrease. Operating budgets become much easier to prepare and are an accurate reflection of actual operating conditions, not the result of educated guesswork.

Capital equipment budgets also become much less of a problem to prepare. The master plan has collected cost data on operations and has also provided an accurate history of repairs and time worked. Use of this data base, together with projected production figures, labor rates, and prices of new equipment, enables the department to make accurate cost justifications to back up their requests for new machinery.

Not only can ownership and operating costs be calculated and compared, but other factors can be considered when making decisions concerning new equipment. Reliability can be established, and this will influence the choice of new makes or models. Armed with accurate historical data, management can establish specifications for new machinery more precisely, so that new units more closely match job requirements.

The preventive maintenance master plan has established organization, methods, and control over equipment and costs. The department staff can concentrate on managing and planning. The maintenance manager and staff then become valuable members of the management team. Their input, together with those of the other departments, will enable the company to operate at a higher level of efficiency and to stay abreast, if not ahead, of the competition.

# chapter 9

# Mechanical Systems Integrity Management

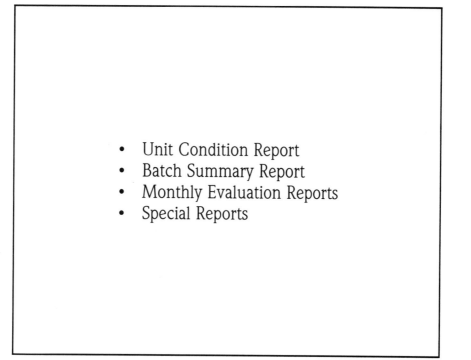

- Unit Condition Report
- Batch Summary Report
- Monthly Evaluation Reports
- Special Reports

Up to this point, the basic structure of a preventive maintenance master plan has been set down in detail. The maintenance department has been organized and a system has been established that provides for routine servicing of the equipment. There is in place a method for collecting cost and repair data on the capital equipment. The maintenance department manager will have available the basic data needed to run the department in an efficient manner and be able to project the requirements of the department and the capital equipment needs on an accurate basis.

Technical advances have placed a heavy burden on maintenance departments. The upkeep and repair of various types of equipment with different technologies and service requirements can be an overwhelming task if it is not organized and controlled. This book has provided the basis for accomplishing this task. At this point, we can carry maintenance management one step further and make use of the latest available technology to help maintain the capital equipment at the highest possible level of operating efficiency. By building on the basic structure of the preventive maintenance master plan, we can fine tune the system to include control over the internal workings of the machinery. This is called *machinery integrity management.* Another description generally applied is *internal on-condition monitoring.*

To get the greatest operational efficiency possible out of the equipment, programs must be put in place to reduce breakdowns, repairs, overservicing, and overmaintenance. This will automatically improve the availability and reliability of the equipment, and the department becomes more fiscally responsible and efficient. We have established a program that goes a long way toward achieving these goals. The routine service and maintenance portion of the preventive maintenance master plan is based on service in accordance with manufacturers' recommended procedures and service intervals. However, maintenance personnel still have little information about the internal status of the machinery. They should be able to determine if a unit is in top internal shape, beginning to show signs of wear, or is deteriorating rapidly, and exactly why a specific situation exists.

By changing oils at prescribed intervals, maintenance personnel attempt to ensure that each unit is receiving the best quality lubrication. What is not known is the exact condition of the oil that is removed at each oil change. The oil may be contaminated or its additive pack may be depleted. On the other hand, it might be in perfect condition. What is needed is a system to analyze the oil and determine its actual condition and give the reasons why it has varied from like-new condition. One of the leading causes of equipment failure can be traced to lubricating oil condition or lack of lubrication.

There are many types of analytical services available. They range from labs that simply analyze individual oil samples to identify 10 or 11 wear metals suspended in the lubricant to full-service mechanical integrity management consulting firms. They all use an oil sample as a starting point, but this is about the only thing they have in common. In the first case, the lab indicates the amount of wear metals present in the sample. This only gives an indication of the state of the lubricating oil when the sample was taken and nothing more. If random sampling is being done with no set schedule, the most that can be expected from this service is an indication of when to replace contaminated oil.

Other analytical labs go further with their services by using spectro-analysis combined with viscosity checks to give a wider range of information. Many maintain records of individual units monitored so they can observe trends in a unit's lubricant condition.

The best of these firms employ the latest techniques in analysis to provide the most information possible concerning the condition of lubricants and the resulting mechanical condition of the machinery being monitored. They make use of atomic emission spectrometers for the detection of suspended wear metal particles, such as iron, silver, lead, tin, aluminum, chromium, copper, magnesium, nickel, and titanium. This same analytic technique will also detect the presence of sodium, barium, calcium, phosphorus, and zinc, which normally derive from the additives in the lubricant. Other elements can be measured, such as silicon and vanadium, which are indicative of contamination. In all, approximately 20 trace elements can be measured, in parts per million, in less than a minute.

Infrared spectrophotometry is also used as a routine analytical technique to determine the chemical composition and contamination level of used oil by comparing it to new oil of the same specification and from the same supplier. Contaminants commonly detected using this technique include water, blowby products, coolant chemicals, unburned fuel, and entrained gases. Degradation of the oil through nitration and oxidation polymerization can be measured directly, as can the depletion of the oil's additive package. This qualitative/quantitative analysis of the differences between used and new oil allows a lubricant to remain in use until it is no longer in like-new condition.

It is thus possible to monitor a combined total of approximately 37 trace elements and contaminants that may be present in the oil. To a full service consulting lab that monitors equipment on a continual basis, each of the elements or contaminants indicates a particular symptom inside the machinery. Viscosity is the third measurement that must be employed to help establish the condition of the oil.

Other checks that are often done are analysis of fuels taken from the client's stock and also analysis of new oils. These checks are done periodically to ensure that clients are getting the quality they expect from their suppliers. These techniques, when combined, will provide the data needed for determining the condition of the equipment from which the samples were taken. The laboratories that are in the vanguard of state-of-the-art lube oil monitoring have progressed beyond the point of mere lube oil analysis. By using a continuous monitoring program, based on a strict schedule, they provide a consulting service of immense value to the client. They do this by looking at the changes occurring in the oil. Most important to the maintenance manager are the results derived from this type of service. The consulting laboratory, using this data base, can not only tell the manager what is happening with the oil but, more important, why it is happening and how the situation can be corrected. The manager does not need to open up a unit to determine its condition, but knows what it is and has complete confidence in the consulting service and the equipment. The manager can see inside without opening units because he or she has been receiving continual reports on each unit and so can make corrections to avoid major problems.

The balance of this discussion dealing with machinery integrity management will be confined to a description of the services provided by these consulting laboratories and how they fit into the overall organization and operation of maintenance systems. Making use of a service such as this extends confidence and control beyond the scope of the basic preventive maintenance master plan.

To take full advantage of the latest technology available to the field of maintenance management makes good sense. It is virtually impossible for maintenance personnel to become expert in all types of equipment, lubrication products, and *tribology*—the science of friction, wear, and lubrication. Modern equipment is just too complicated, sophisticated, and expensive to leave to chance any phase of maintenance.

Machinery monitoring laboratories that provide consulting services generally do so on an annual contract basis, as opposed to charging by the sample. This is one way to differentiate between simple oil analysis and full-service consulting firms. In the long view, the consulting firms cost less and provide greater savings, as will be illustrated.

Another important factor to consider when discussing maintenance management is the quality of lubricants and fuels. Not many managers give much thought to quality beyond ensuring that they are using the specified types and grades for their equipment. When purchasing lubricating oils from reputable oil companies, they assume that the products always meet specifications. This is not always the case. There are often mistakes caused by human error and other factors. The same type and grade of oil may vary considerably, depending on where the product was produced. One plant may have better quality control than another. Oil drums have been known to be mislabeled, or bulk deliveries arrive with the wrong product. It is thus possible for a user to discover that he has hydraulic or turbine oil in the crankcases of his diesel engines.

Fuel quality varies considerably. The origin of the crude stock has a great influence on the quality of the fuel made from that stock. The sulfur content of diesel fuel has a notable effect on the operation and maintenance of engines. A consulting service will also monitor newly received lubricants and fuels to ensure that they meet set standards. In the case of substandard fuels, the lab can offer suggestions on how to compensate for their adverse characteristics.

The use of infrared spectrophotometry is essential to this type of service in that it will immediately indicate any changes in the new lubricant compared to the same lubricant known to be of standard specifications. Even if a sample of new oil has not been received for a long period, the infrared analysis of a used oil sample may indicate that something is different with the oil currently in use. Since oil monitoring is done on a systematic basis, the consulting lab has a broad data base on each unit and is watching trends in the condition of the oil. If the normal trend, usually a slight degradation, should drastically change, the lab will question the results. If the change does not fall into the normal pattern, a resample will be requested along with a sample of unused oil. The unused oil sample will be treated in the same manner as the used oil sample. Since the infrared analysis is based on a direct comparison of an oil sample to unused oil that is known to be to correct specifications, any change in the quality or makeup of the new oil will be identified. This service ensures that clients always have the correct oil in their units. Viscosity checks are also done to help determine if the lubricants and fuels are meeting the required specifications.

Fuel must be monitored continually because its quality affects the condition of engines and lubricating oils. There may be little that can be done to change the quality of fuel on hand, or being delivered, but steps can be taken to alleviate the problems associated with low-quality fuels.

One of the most common problems is a high-sulfur content in diesel fuel. When burned in an engine, the by-products of combustion cause acidity to build up in the lube oil and accelerated component wear. The acidity can be combated by using a lubricating oil with a higher total base number (TBM). The increased wear is caused more, however, by the ash released into the system as the result of combustion. This is of more immediate concern than the much slower action of increased acidity. Ash acts like a grinding compound on mating engine parts. The problem is often made worse by the lubricating oil itself. Those lubricants having a high TBM also have a high ash content. This ash, combined with that present as a by-product of combustion, compound the problem of engine wear.

There are ways to control problems associated with low-quality fuel. Generally, a consulting laboratory will suggest a middle-of-the-road approach that will provide the optimum engine protection consistent with reasonable cost. One course of action that was successfully implemented to overcome the problems associated with high-sulfur diesel fuel was recommended by a client's consulting lab. After studying past data and analyzing the fuel, it was recommended that the lubricating oil be changed to one with a slightly higher total base number. This would help somewhat to overcome the acidity buildup but not to the point where the oil would add to the existing ash problem. Second, kerosine was added to the diesel fuel so that a 50–50 mixture was used in the engines. This effectively reduced the amount of sulfur per unit volume and improved combustion.

Another way a full-service consulting firm can be of assistance to a client is to monitor the results obtained when running selected test units operating on different lubricants. Sometimes a different oil may provide better, or the same quality, lubrication at a lower cost. Sometimes, the application of a unit, or the environment in which it is working, may indicate that a change is required from the lubricant normally used in that machine. The lab can help in determining the best lubricant to be used. Then there is the situation where the wrong lubricant is inadvertently used. A case in point is a company using diesel-engined tractors, trucks, and irrigation pumps. The company standardized on one high-quality engine oil for the entire fleet. Mixed in were some irrigation pumps powered by air-cooled engines using natural gas for fuel. The oil usage rate was high on these units. They were using the same engine oil as the rest of the fleet.

After a week of continuous use, the oil would thicken to the point where personnel became concerned and would change it. The problem was isolated by the consulting lab and found to be one of polymerization caused by the spark ignition used on these engines and the low engine operating temperatures. Knowing this, a different oil was selected for these engines and the problem was eliminated. Lubrication was improved and engine oil changes went from approximately 168 hours to over 600 hours. A check with the engineering department of the engine manufacturer resulted in a recommendation to use the same oil as the lab had selected. The company had mistakenly been using the wrong oil in these particular units. Confidence in the consulting laboratory was reinforced as a result of this one incident.

Continuous mechanical integrity monitoring has the advantage of eliminating the costly practice of overmaintenance. The more conscientious the maintenance department, the greater the inclination toward overmaintaining the equipment. This is not a case of bad judgment on the part of the manage and staff. It is usually the result of following manufacturers' recommendations to the letter so as not to jeopardize the warranty or the integrity of the machine. They may also want to be on the safe, side since they are not sure of the internal state of the equipment.

Most manufacturers of machinery tend to be very conservative when recommending periodic services for their machines. This is particularly true when dealing with engines and the intervals set for oil and filter changes. Common practice is to call for an oil and filter change every 200 to 250 hours for normal working conditions. One of the problems facing the maintenance staff is how to determine what is normal. A second problem is the cost of operating a unit or a fleet of units. If oil must be changed at reduced time intervals, the cost of operation increases. Not only must more oil and filters be used per given time period but the unit must be taken out of production to perform this maintenance. This not only reduces available production time but the service costs increase as well.

One major producer of agricultural and construction equipment recommends changing oil and filters every 50 hours if the sulfur content of the diesel fuel exceeds a certain level, common in many areas of the world. With no other course of action open to prospective owners, they may decide that they cannot afford to own the machine under these circumstances.

The reason manufacturers tend to be very conservative is that they have little control over the machine once it is in the hands of the owner. Faced with the prospect of absorbing high warranty costs, they place stringent service and maintenance procedures on their products. If the owner follows these recommendations, warranty claims will be reduced and the maker can be reasonably sure that those claims that are received are valid. In most cases, this is a reasonable approach. Sometimes, recommendations can approach extremes, as in the case sited above. In fairness to the manufacturers, some of the treatment given new equipment would make a grown man cry. It has been said that "a machine will last forever if it is not touched by human hands." Unfortunately, there is more than a little truth in that statement. Equipment manufacturers are not about to consider warranty claims where damage has resulted from ignorance or carelessness.

There are thousands of examples of poor service and maintenance practices that could be listed, but space and time prohibit any detailed discussion. In general, they all fall into the following broad categories: lack of cooling system maintenance, use of improper lubricants, failure to maintain electrical systems, poor training of personnel, and little or no maintenance organization. It is interesting to note that those maintenance organizations that strive to do a creditable job are the ones who are most eager to learn how to increase their effectiveness. They are the ones that adapt machinery integrity management, based on continuous lube oil monitoring, into their systems. Once such a system is in place, the results benefit both the owner and the manufacturer. One realizes greater productivity at lower cost and the other receives fewer claims against its product. Machinery manufacturers become quite enthusiastic about this approach to maintenance once they observe it functioning and see the results obtained from a well-run program.

Before the use of infrared spectroanalysis became an integral part of the routine monitoring procedure, it was very difficult to demonstrate, in advance, tangible savings associated with a machinery integrity program. Savings were easily demonstrated at the end of the first year in a program by comparing repair costs and downtime to those of previous years. These savings vary with individual organizations and obviously are difficult to predict prior to installing the system. With infrared spectroanalysis assessing the actual condition of the oil in a unit, the oil can be used until it is shown to have deteriorated to a point where it must be replaced. Data accumulated on this subject reveal that oil and filter change intervals can be extended an average of two to three times the normal recommended intervals. If the conservative figure of doubling the interval is used, that in itself is usually enough to demonstrate the economics of adopting the program. The other savings attributed to the program usually represent pure profit to the client.

To make this program work, the maintenance group must ensure that routine checks and services are carried out on a strict schedule as outlined in the preceding chapters. Lube oil levels must be maintained and personnel must ensure that the equipment is kept in good working condition. Oil sampling must be done on schedule. If a problem is suspected with a unit, it should be sampled immediately and the consulting lab advised as to the reasons for the special sample. It is a two-way street between client and consulting lab. The more information flowing between the two increases the control over the condition of equipment and also the amount of mutual trust built up between client and consultant.

One point should be made here about extending the use of lubricating oils. By constant monitoring of the oil's condition, there have been cases where units have had their change intervals extended to over 1500 hours and some even to 3000 hours. These are exceptional cases where the client company maintains strict control over their service and maintenance. Many maintenance people have a negative reaction to these statements and the comment most often made is: "You can't run oil that long in an engine."

The answer to that is simple. The same oil was not used throughout those total hours. There is always makeup oil to be added. All engines use oil, no matter how good their condition. The additive package of the oil will gradually deplete but makeup oil tends to reinforce it. Thus, at the end of 3000 hours, the oil in the crankcase is not the same oil as was present at the start. Remember, lube oil deteriorates because of contamination and degeneration of the additive package. Excess heat can also cause changes, as can the products of combustion, but all of these problems can be detected and identified long before equipment damage occurs. These long oil change intervals can be attained only by the strictest maintenance of machinery and are admittedly examples of the extreme end of the scale. Air intake systems and fuel injection systems must be kept in peak condition. Fuel and lubricants must be correct for the machinery and conditions. The consulting service is not a substitute for good maintenance practices. It is an additional management tool.

Since wear is always present in machinery, repairs will have to be made from time to time. With a machinery integrity program these repairs can be reduced in frequency and scheduled well in advance, so when they are done, less time and money are expended. The machinery repairs can be scheduled so that losses in production time are kept to a minimum. In addition, the work required will usually be less because the problem is known beforehand, so minor problems do not have a chance to develop into major breakdowns. The costliest repairs are those resulting from unforeseen breakdowns. Administered properly, machinery integrity management programs go a long way toward eliminating catastrophic breakdowns.

Operating an effective program requires only a limited amount of work over what is required for the basic preventive maintenance master plan. In Chapter 10 a step-by-step analysis is made of how such a monitoring program is set up and operated in a typical maintenance department and what services the consulting lab normally provides.

The maintenance department and management need certain information from the lab in order to gain the most from the monitoring program. Information forwarded to the client company is in the form of reports, some of which are directed to the maintenance department and the people intimately involved with the day-to-day service of equipment. Other reports are sent to management to provide an overview of the entire program.

## UNIT CONDITION REPORT

The unit condition report is generated as soon as an oil sample has been completely analyzed by the consulting lab. The report is sent to the client for each component sampled and contains that component's history and current analytical results. It indicates any problems and recommends specific corrective actions to be taken by maintenance personnel. Based on the analysis, the current condition of each component is reported as being in one of three possible categories: normal, borderline, or critical. Any discrepancies or deficiencies in either mechanical or lubricant condition are identified in lay terms, as is the remedial action specified. Critical reports are also sent in by telephone or telex communication so as not to lose valuable time.

## BATCH SUMMARY REPORT

The batch summary report summarizes results from each group of samples received by the lab. In normal operations, samples are submitted to the lab in quantities ranging from 10 to 25. The size of each submission, or batch, depends on the size of the client company and the method used to transmit the oil samples. This summation is for the use of maintenance supervisors and focuses administrative attention by providing a continuous update of equipment status, units requiring maintenance action, and units due for sampling. The supervisor has an overview of how the system is working and the effectiveness of personnel in carrying out the program.

## MONTHLY EVALUATION REPORTS

Monthly evaluation reports are for higher management levels. They integrate monitor data and program information into a management-oriented overview of the mechanical integrity of the operation and the effectiveness of maintenance procedures being applied. The emphasis is on highlighting trends and identifying problem areas. This is done without burdening the managers with a lot of low-level details. They do provide the information required to identify and remedy problem areas. They also summarize the results of the program in two sections.

The first section is presented in graphic form for the month, year to date, and prior year. It measures the effectiveness of operations in terms of equipment condition and supervisory control. The second section consists of detailed tabulations showing unit status, unit condition changes, units needing special attention, and critical unit discrepancies. This section is also presented in cumulative form and can be set up by divisions. The latter type of format depends on the size of the client and how its management wants to group the equipment for reporting and comparison purposes.

## SPECIAL REPORTS

Information generated by a program of mechanical systems integrity management can serve as a basis for a wide variety of special studies, such as comparative evaluations of equipment, lubricants, filters and filtering systems, mechanical components, maintenance, and operational procedures. After a program has been in operation for a sufficient length of time to accumulate a solid data base, these reports and studies can be made with a high degree of accuracy. They are of tremendous help to management because they present data concerning equipment and operations that might otherwise never be known. Once management is made aware of problem areas that cause reduced operating efficiency and higher costs, it can take steps to eliminate or reduce the situation. The most common types of special reports are listed next.

1. *Statistical wear profiles.* These present the frequency with which certain wear metals and pollutants appear in a component over an extended time period.
2. *Confidence factor analysis.* This study ranks equipment according to the level of confidence that can be placed in it, specifically in the condition of the internal parts. It is presented in both graphic and tabular form so that individual units or groups of units can be identified and rated.

These two special reports help form the basis for other special studies and analyses done at the request of the client company. For instance, the client may want to have a study conducted to determine the possibility of changing engine oil to a less costly but better lubricant for its particular operating conditions. Another client might wish to run a comparative study of two groups of equipment, distinguished by make or model, to determine which type is most suited to a particular operation. With a good data base to work from, the consulting lab can act as a partner and technical backup for the client company and assist in making decisions concerning equipment, maintenance procedures, procurement, and management techniques.

Essentially, this is an advanced maintenance management program based on state-of-the-art technology and an on-condition maintenance philosophy. A program, as described here, removes the uncertainty concerning the reliability and internal condition of the machinery being operated. Management can turn its attention to providing overall supervision of maintenance, planning, and scheduling predetermined minor repairs.

In Chapter 10 a step-by-step procedure is described for integrating a program of mechanical systems integrity management into the basic system already established. At the same time, examples of the setup by the client and reports from the consulting lab will be illustrated. The addition of this type of management program can change the basic concept of maintenance from one of repair to one of planned management.

Once a company decides to use a monitoring program, there are a number of decisions that must be made at the start concerning the scope of the service and coverage. At this time, the client company will discuss its needs with a representative of the consulting lab. The first steps toward implementing a program will be agreed to and the scope of the monitoring program will be established. All this will vary according to the size and type of business. One client company may wish to monitor some components because of special circumstances or working conditions. Another company might not bother monitoring the same components, except on an extended basis. Then there are special cases that may require closer surveillance than normal, as is the case with small aircraft engines.

One of the first things to establish is the sampling interval for each component or machine. This can be done in intervals of hours, weeks, or months. It can also be based on the amount of fuel consumed or distance traveled. Once the number of components to be monitored is established and sampling intervals set, the company will have a good idea of the amount of time and the number of personnel needed to run the program.

## PERSONNEL

The job of implementing and administering a monitoring program should be assigned to the service manager, who will have overall responsibility for seeing that the program is carried out in an efficient and thorough manner and that all recommendations of the consulting lab are implemented on schedule. The service manager will also be the person who is in closest contact with the lab since he or she will be discussing problems, remedial actions, and special studies on a routine basis. See Chapter 3 for the organization of the service group and personnel requirements. Refer also to the end of Chapter 2 and the basic organization chart for a maintenance department.

As soon as the scope of the monitoring program has been established, it will become evident what personnel will be needed to perform the day-to-day operations. The type of industry will also be a factor in determining how many people will be assigned to the monitoring program. A large agribusiness, with mobile equipment spread over a wide area, will use more people than a power-generating plant, where the equipment is stationary and limited to a relatively small area.

At the very least, there must be a clerk to handle scheduling, paperwork, and sample shipments, as well as a trained technician to do the actual sampling of equipment components. In small companies, these two functions can be accomplished by one person. Much larger companies may use two or more clerks and as many sampling technicians. These people report directly to the service supervisor, who in turn reports to, and counsels, the service manager.

The clerk's position should be filled by a person who is skilled in the art of organizing work and keeping track of large numbers of equipment and samples. The clerk must have a basic grasp of mechanical equipment. The same characteristics are required of the technicians in charge of sampling components. Each must have enough mechanical know-how to appreciate the importance of the work and to understand what the machinery components are and how they fit into the program. It is not necessary that they come from the ranks of mechanics or foremen. It is necessary that they be hardworking and conscientious. This trait can be reinforced if the people assigned to the program are made aware of the hows and whys at the very start. Once they understand the importance of the work and become true believers, the rest is easy. Experience shows that well-trained clerks and technicians, who understand the benefits of a monitoring program, will be its biggest boosters and will perform effectively with a minimum of direct supervision. Many times, they are the ones who will advise their superiors of abnormal situations and make suggestions for correcting them. The consulting lab will help train and motivate these people.

## MONITOR CONTROL LIST

The service supervisor and the newly assigned clerk must first make a list of each component that will be monitored in this program. This is good training for them in the administration of the program because they will become familiar, from the start, with all the equipment to be included. The equipment inventory list, described in Chapter 4, is used as the source of information in creating the monitor control list. An example of a monitor control list is shown in Fig. 10.1, which shows the first page of a multipage list.

The company number, description, and model are listed and taken directly from the equipment inventory list. Additional columns are set up for the monitor number, component, and lubricant used in that component. Monitor numbers are usually listed in order from the first down through the last component, starting with the number 1. The unit column now lists the description of the machine, and the component column lists the component to be monitored. Another difference between the two lists is that one unit may have four or five monitor numbers. They would denote different components on the same unit. These might include the engine, transmission, and front and rear differentials.

The monitor list is made by hand and submitted to the consulting lab. Their personnel enter this information into a computer as the first step in establishing a program for a new client. The format of the list can be established by the client company. One company might wish to group the list by equipment type, component type, or perhaps by components using the same lubricant. Another method is to make separate lists for each company division. This is useful for comparative purposes when monthly and special reports are sent to management. Once this information is entered

```
---------------------------------------------------------------------------
----------- SPECTRON  INTERNATIONAL,  INC. -------------
---------- MECHANICAL  INTEGRITY  ANALYSIS --------------
-----------------------------------------------------------Copyright 1988 Spectron
---------------------------------------------------------------------------

                    AGRICULTURAL (SUGAR) OPERATION
                         MONITOR CONTROL LIST

REVISION DATE: 27 APR 1989   (0)
                              WATER CONTROL

 MON. NO.    CLIENT ID.    UNIT DESCRIPTION    COMPONENT      MODEL        LUBRICANT

      1    5001         WATER PUMP          ENGINE      DDA 4-53       TEXACO URSA SP-30
      3    5003         WATER PUMP          ENGINE      DDA 4-71       TEXACO URSA SP-30
      5    5005         WATER PUMP          ENGINE      CAT D-318      TEXACO URSA SP-30
      6    5006         WATER PUMP          ENGINE      CAT 3304       TEXACO URSA SP-30
      7    5007         WATER PUMP          ENGINE      DDA 3-71       TEXACO URSA SP-30
      8    5008         WATER PUMP          ENGINE      DDA 3-71       TEXACO URSA SP-30
      9    5009         WATER PUMP          ENGINE      DEUTZ AF4L912  TEXACO URSA SP-30
     10    5010         WATER PUMP          ENGINE      DDA 3-71       TEXACO URSA SP-30
     11    5011         WATER PUMP          ENGINE      DDA 4-71       TEXACO URSA SP-30
     13    5013         WATER PUMP          ENGINE      DDA 4-71       TEXACO URSA SP-30
     14    5014         WATER PUMP          ENGINE      CAT D-315      TEXACO URSA SP-30
     15    5015         WATER PUMP          ENGINE      DDA 4-71       TEXACO URSA SP-30
     16    5016         WATER PUMP          ENGINE      DDA 4-53       TEXACO URSA SP-30
     17    5017         WATER PUMP          ENGINE      DDA 4-71       TEXACO URSA SP-30
     18    5018         WATER PUMP          ENGINE      CAT D-315      TEXACO URSA SP-30
     20    5020         WATER PUMP          ENGINE      DDA 3-71       TEXACO URSA SP-30
     21    5021         WATER PUMP          ENGINE      DDA 3-71       TEXACO URSA SP-30
     25    5025         WATER PUMP          ENGINE      DDA 4-71       TEXACO URSA SP-30
     30    5030         WATER PUMP          ENGINE      DDA 4-71       TEXACO URSA SP-30
     31    5031         WATER PUMP          ENGINE      DDA 4-71       TEXACO URSA SP-30
     32    5032         WATER PUMP          ENGINE      DDA 4-71       TEXACO URSA SP-30
     33    5033         WATER PUMP          ENGINE      DDA 4-53       TEXACO URSA SP-30
     34    5034         WATER PUMP          ENGINE      CAT D-315      TEXACO URSA SP-30
     35    5035         WATER PUMP          ENGINE      CAT D-330      TEXACO URSA SP-30
     36    5036         WATER PUMP          ENGINE      CAT D-330      TEXACO URSA SP-30
     38    5038         WATER PUMP          ENGINE      CAT D-315      TEXACO URSA SP-30
     39    5039         WATER PUMP          ENGINE      DDA 4-71       TEXACO URSA SP-30
     40    5040         WATER PUMP          ENGINE      CAT D-330      TEXACO URSA SP-30
     44    5044         WATER PUMP          ENGINE      JD 6466DF      TEXACO URSA SP-30
     45    5045         WATER PUMP          ENGINE      CAT D-330      TEXACO URSA SP-30
     46    5046         WATER PUMP          ENGINE      CAT D-318      TEXACO URSA SP-30
     47    5047         WATER PUMP          ENGINE      DDA 6-71       TEXACO URSA SP-30
     48    5048         WATER PUMP          ENGINE      CAT D-330      TEXACO URSA SP-30
     49    5049         WATER PUMP          ENGINE      CAT D-330      TEXACO URSA SP-30
     50    5050         WATER PUMP          ENGINE      DDA 4-71       TEXACO URSA SP-30
     51    5051         WATER PUMP          ENGINE      AC 10000       TEXACO URSA SP-30
     52    5052         WATER PUMP          ENGINE      CAT D-333      TEXACO URSA SP-30
     53    5053         WATER PUMP          ENGINE      CAT D-330C     TEXACO URSA SP-30
     55    5055         WATER PUMP          ENGINE      DDA 4-71       TEXACO URSA SP-30
     56    5056         WATER PUMP          ENGINE      DDA 4-71       TEXACO URSA SP-30
     57    5057         WATER PUMP          ENGINE      DDA 4-71       TEXACO URSA SP-30
     58    5058         WATER PUMP          ENGINE      DDA 4-53       TEXACO URSA SP-30
     59    5059         WATER PUMP          ENGINE      DDA 4-71       TEXACO URSA SP-30
     60    5060         WATER PUMP          ENGINE      DEUTZ F3L912   TEXACO URSA SP-30
     61    5061         WATER PUMP          ENGINE      DDA 4-53       TEXACO URSA SP-30
     62    5062         WATER PUMP          ENGINE      DDA 4-71       TEXACO URSA SP-30
     63    5063         WATER PUMP          ENGINE      CAT D-333      TEXACO URSA SP-30
     64    5064         WATER PUMP          ENGINE      DDA 6-71       TEXACO URSA SP-30
     65    5065         WATER PUMP          ENGINE      DEUTZ F3L912   TEXACO URSA SP-30
     66    5066         WATER PUMP          ENGINE      DDA 4-71       TEXACO URSA SP-30
```

**Figure 10.1** Monitor control list. (Courtesy of Spectron International, Inc.)

into the computer, the lab will issue a printout to the client and this becomes the master monitor control list. Changes will be made as a normal result of equipment status changes. It is the responsibility of the client company to ensure that the list is kept current.

As soon as a new piece of equipment is put into service, it should be added to the monitor list. This is usually done by breaking out the components to be monitored and adding them at the end of the list. Units that have been permanently or temporarily taken out of service must be removed from the list. It is important to keep the lab updated as to the status of machinery in the program. In a computerized system, the monitor number is the element that triggers reports. If the monitor list is not kept current, the lab will consider any inactive numbers as not being serviced and will periodically show these as neglected units in the management reports. Of course, management will soon discover that the service group is not doing its job when they investigate these "neglected" units and find that they have been sold or inactivated.

Some client companies affix the monitor numbers to components, others do not. It usually depends on the type and number of monitored components on a piece of machinery. On mobile equipment that requires engine, transmission, final drive, and hydraulic monitoring, individual labeling is not normally done. The main reason is that one would not normally be able to see all the labels because of component location or the probability of dirt and grime covering the component. In this case, the clerk posts the latest copy of the monitor control list and simply cross references to label the sample bottle. If the transmission of the unit with company number 518 is to be sampled, he lists it that way for the technician who will take the oil sample. The sample bottle comes in with the label reading "518 -Trans." and the clerk then adds the monitor number to the bottle label. The lab is only interested in getting the correct monitor number on the sample bottle, but any additional information certainly does not hurt.

## STATUS BOARDS

Chapter 6 contains a discussion of the use of status boards and how they are used in the basic preventive maintenance master plan. Figures 6.1 and 6.2 illustrate the boards and the index cards. The same type of status board should be used in the monitoring program. This display is primarily for the use of the clerks and technicians who are working with the program. It gives them a handy visual display so they will know when to schedule units and/or components for sampling.

Do *not* use the same status boards for the monitoring program as those used for routine lubrication and maintenance service. However, flags can be put on these boards to indicate that an oil sample has been taken. This should be left to the discretion of the service manager and supervisor. The boards used for routine service and maintenance list all equipment by company number, the same way they are listed on the equipment inventory list. These are all complete units, not components as are listed on the monitor control list and monitor control status boards. Another difference is that they may have different intervals for service than for monitoring. It is possible to have service based on hours of operation or fuel consumed and the monitoring program based on weeks and months. In smaller companies, the same clerks can run both functions but they must be handled separately.

The components should be listed in the order in which they appear on the monitor control list. An index card with the monitor number, company number (of the complete unit), component name, and model is to be filled out for each component. The status board should then be set up for the type of monitoring intervals that have been established. Normally, the intervals are based on elapsed time, in which case the boards might cover a 12-month period. They can also be set up on the basis of the amount of fuel consumed. On the line next to each component, a colored peg can be set at each interval at which a sample is to be taken. As the string peg approaches one of these colored markers, the clerk is made aware that a sample needs to be taken and can schedule these on a timely basis so that the sampling technicians can perform their tasks efficiently.

## ESTABLISHING MONITORING INTERVALS

The client and consulting lab established monitoring intervals as one of the first steps in implementing the new program. As stated earlier, intervals are usually based on time, either by month or by weeks. This is a much easier method for service personnel since the monitor list covers only working units. Units that are out of service, or on standby, are not on the active list.

If the client company has been giving routine service and maintenance on the basis of hours, the normal periods for oil changes can easily be converted to the nearest number of weeks or months. Simply determine the average number of hours worked per week and apply this to the normal oil change interval. As a rule, oil sampling is done in place of the normal oil change. Assume that a crawler tractor has its engine oil and filters changed every 200 hours. If the unit is normally working eight hours per day, six days a week, or 48 hours per week, that means it will have worked about 206 hours during the course of a month. In this case the unit will be sampled once a month while on active status.

The same reasoning can be used if that unit was being serviced on the basis of gallons of fuel consumed. In cases where the client or consulting lab feels the interval is not realistic, it can be decreased or expanded. The best approach is to start off on the conservative side. The client may have some units that are troublesome. These should be monitored at shorter intervals to try to isolate the problem as soon as possible and quickly build up a good data base for trend analysis.

Transmissions and gearboxes may have longer monitor intervals. Final drives may go up to three months between samplings. It has been found that field tractors whose differentials are checked too frequently have more problems than those on extended intervals. This is because dust and dirt find their way in through the filler plug each time it is removed. However, troublesome units should be monitored more frequently until the cause of a problem is determined. Each component to be monitored must be studied individually when setting up the initial sampling intervals.

At the end of the year, a reappraisal can be made and the intervals adjusted as required. When a component shows signs of abnormalities, the lab will automatically ask the client to sample on a shorter interval until the situation is brought back to normal. Clients who feel unsure of a component can sample at any time and advise the lab of the reason for an unscheduled sample. Since consulting labs charge a flat fee by the month, or year, varying amounts of oil samples will not change the cost to clients. The most important point to remember is that lab and client should work together as a team.

## TAKING AND SUBMITTING OIL SAMPLES

Once the status boards are in place and set up to show the monitoring intervals, the taking of samples can begin. When starting a new program, all components should be sampled. Do not make the mistake of trying to sample them all at the same time, or even all in the first week. If this is done, there will be another rush to collect samples as the next monitoring interval approaches. Plan the monitoring routine so that sampling is spaced out over approximately one month. If the program is initiated at the beginning of a month, the last components to be monitored should be sampled just before month's end. This will ensure that a more even distribution of work and sample flow is established and will continue during the months and years ahead. Another reason for this approach is the fact that the lab will often request that resamples be taken to check on questionable components. The client company may also initiate special sampling. These samples will add to the normal sampling work load. It pays to start off with a well-planned schedule because that will be the one that must be adhered to for the life of the program.

Each day, the clerk ensures that the sampling technician is supplied with a list of components to be sampled. It is also a good idea to sample the stock of new oils and fuel once each month. These should be labeled by brand and designation: for example, "Shell Rotella T, SAE 30, Stock Sample." These new samples will provide the lab with the correct reference oil against which used oil samples will be compared during the process of analysis. The lab can also check to see that the new oil and fuel samples are to specifications. Fuel will be checked for quality because its combustion greatly affects the condition of engines and lubricating oils.

Samples are not usually sent to the lab individually, except in special cases. The normal procedure is to send samples in lots, or batches, to save space and shipping costs. The distance from the lab and method of shipment will determine the batch size. Air shipment is a common method used where distances are long. Small cartons containing about 20 sample bottles are the most common type used for air shipment. The clerk who receives the oil samples makes sure that each bottle is labeled with the correct monitor number. The clerk also flags the status boards with colored indicators showing that these components have been sampled.

The clerk next enters the monitored sample data on a packing list (Fig. 10.2) that will be included with the batch of oil samples. The name of the company and division, if applicable, is entered at the top along with the

# spectron

## BATCH PACKING LIST

FROM: _____

DIVISION: _____

BATCH NO.: _____

DATE: _____

SEND TO:

SPECTRON INTERNATIONAL
COND. LAS LOMAS
PROFESSIONAL CENTER #3
CAPARRA HEIGHTS, PR 00921

| MONITOR NUMBER | UNIT DESCRIPTION | EQUIPMENT MILES/HOURS | LUBRICANT MILES/HOURS | MAKE-UP OIL ADDED | REMARKS |
|---|---|---|---|---|---|
| | | | | | |

**Figure 10.2**   Batch packing list. (Courtesy of Spectron International, Inc.)

date. The batch number is simply a consecutive number. The first batch would have the number 1. Another method is to add the year so that the first batch might have the number 1–89.

Many clients use only columns 1 and 2 and occasionally the remarks column for flagging specials. This will not detract from the program, but if the other columns are completed, it adds to the overall data base for that component. Equipment hours are an indication of the working age of the component. Lubricant hours are taken to mean the time that has elapsed since the last oil change, and they help in determining the actual oil change interval for that component. As the time on the program increases, these intervals will increase to a certain point and then remain fairly consistant. This will depend on the quality of the routine service and maintenance given to the equipment. Noting the amount of makeup oil added between sample monitoring provides the lab with another piece of data that will influence the interpretation of trends showing up in the monitored oil.

If an engine shows a steady decline in the quality of the lubricating oil, and reports from the client show a greater than normal amount of makeup oil being used, this would indicated the presence of a serious problem. On the other hand, if this unit were using a normal amount of makeup oil, the results would be interpreted differently. It pays to fill in as much of this information as possible, even though it may mean a little extra work.

## PERIODIC SERVICE SHEETS

Chapter 6 includes a detailed description of the format and use of periodic service sheets. These are vital tools for maintaining an effective maintenance system. Their use in the basic preventive maintenance master plan remains unchanged with the addition of a lube oil monitoring system. Only the notations calling for an oil and filter change are affected. Figure 10.3 shows the change made on these sheets. This is the sheet for 250-hour service and is the same as that shown in Fig. 6.5 except that the line calling for engine oil and filter change now reads "Sample Engine Oil." It should be pointed out that this line is now only an informational statement and not a work description for the service personnel.

The people giving routine service and lubrication maintenance are not normally the same technicians that have the task of taking oil samples. It is for this reason that some companies prefer to delete the line calling for oil and filter changes without adding further notations. Whichever system is used must be applied universally to all sheets that contain an oil and filter change statement.

## REPORTS AND OTHER INFORMATION

On the following pages we describe the various reports and studies that are routinely made available to a client company.

<u>Unit No.</u>

<u>(MAKE OF) TRACTOR</u>                    <u>(MODEL)</u>

250 HOUR SERVICE

LUBRICATION

1. Oscillation supports on front axle (2)      MPG

2. Mid-ship support bearing (1)                MPG

3. Steering cylinder pivots (4)                MPG

4. Engine fan drive pulley bearing (1)         MPG

5. U-Joints and shafts (10)                    MPG

6. Axles and bearings (4)                       MPG

7. Valve handle bushings (1 to 6)              MPG

SERVICE

1. SAMPLE ENGINE OIL.

2. Drain water and sediment from fuel tank.

3. Check oil level in front/rear differentials. SAE 30

Date service completed:           Signed:

**Figure 10.3**  Periodic service sheet: 250-Hour.

### Unit Condition Report

The unit condition report combines analytical results of the latest sample received and the component's past history to indicate the mechanical condition of the component and its lubricant. Specific maintenance recommendations are made when problems are detected. An example of a unit condition report is shown in Fig. 10.4 and 10.5.

The monitor number is shown at the top of the page together with other pertinent information about the component and its lubricant. Page 1 also lists, at the top, the sample date, hours on the component, lubricant hours, and amount of makeup oil used. The latter three items can be listed only if the information has been included on the batch packing list sent by the client. The latest sample results are always listed in the left column. The others show previous analytical results on the same component. Even though a column of these numbers, by themselves, may mean little to the client, trends can be seen by observing changes in these values from sample to sample. Asterisks denote those values that are outside the norm.

At the bottom of these columns is a line labeled "Results." The number refers to a number coded index that gives a short title to the discrepancy and the letter refers to the color code of the report. The unit condition reports classify the current condition of the component as *critical* (red), *borderline* (yellow), or *normal* (white). Those reports classified as critical are flagged with a red sticker, borderline reports will have a yellow sticker, and normal reports are not flagged. Critical reports are also passed on to the client by telephone, telex, or telegram so that the client can take immediate action without having to wait for the report to arrive in the mail.

The discrepancy is stated at the bottom of the page together with the recommended corrective action. The second page is a repeat of this information and is meant for use by the service personnel as a worksheet. The service clerk will issue this sheet to the people giving routine service and maintenance. They will use it as their worksheet and carry out the instructions noted if action has not already been taken. These reports are for the use of the service department supervisor and personnel who are directly involved in the day-to-day service of the equipment.

### Batch Summary Report

The batch summary report summarizes the results of each batch of oil samples. It is aimed at the next-higher level of management. In this case it would be the service manager, maintenance manager, or both. The report provides the staff with a continuing update of equipment status, required maintenance action, and the equipment due for sampling. This is simply a condensed accumulation of the most important data taken from the unit condition reports of this batch. It provides the managers with a concise overview of the status of the equipment and how the service personnel are performing their duties without burdening management with technical details. Management thus has the necessary information to control the routine maintenance process.

Figures 10.6 through 10.11 illustrate a typical batch summary report.

```
------------------------------------------------------------------------------------
----------  S P E C T R O N   I N T E R N A T I O N A L ,   I N C .  --------------
----------  M E C H A N I C A L   I N T E G R I T Y   A N A L Y S I S  -----------
------------------------------------------------------------Copyright 1988 Spectron
                  UNIT CONDITION REPORT: MONITOR NO. 228                  REV: 1
```

CLIENT: AGRICULTURAL (SUGAR) OPERATION                          DIVISION: TRACTORS
CLIENT ID.: 2050              UNIT DESCRIPTION: CASE TRACTOR    COMPONENT: ENGINE
MODEL: CASE 1896             LUBRICANT: TEXACO URSA SP-30

| | | 27 APR 89 | 27 MAR 89 | 24 FEB 89 | 30 JAN 89 | 10 JAN 89 |
|---|---|---|---|---|---|---|
| SAMPLE DATE: | | 27 APR 89 | 27 MAR 89 | 24 FEB 89 | 30 JAN 89 | 10 JAN 89 |
| EQPT HOURS: | | N/P | N/P | N/P | N/P | N/P |
| LUB HOURS: | | N/P | N/P | N/P | N/P | N/P |
| MAKEUP LUB: | | N/P | N/P | N/P | N/P | N/P |
| TRACE | Iron | 99 | 27 | 74 | 31 | 16 |
| ELEMENTS | Silver | 0 | 0 | 0 | 0 | 0 |
| PPM | Aluminum | 12 | 3 | 8 | 3 | 3 |
| | Calcium | 3370 | 3250 | 3890 | 2985 | 2330 |
| | Chromium | 10 * | 2 | 5 | 2 | 1 |
| | Copper | 4 | 2 | 6 | 3 | 1 |
| | Magnesium | 4 | 3 | 4 | 3 | 3 |
| | Sodium | 0 | 24 | 27 | 20 | 18 |
| | Nickel | 0 | 0 | 1 | 0 | 0 |
| | Lead | 9 | 4 | 16 | 4 | 3 |
| | Silicon | 7 | 6 | 7 | 4 | 0 |
| | Tin | 0 | 0 | 0 | 0 | 0 |
| | Titanium | 0 | 0 | 0 | 0 | 0 |
| | Boron | 0 | 0 | 3 | 2 | 0 |
| | Barium | 0 | 0 | 0 | 0 | 0 |
| | Phosphorus | 1850 | 1530 | 2005 | 1861 | 1500 |
| | Manganese | 2 | 1 | 1 | 2 | 0 |
| | Molybdenum | 0 | 1 | 0 | 0 | 0 |
| | Vanadium | 0 | 0 | 0 | 1 | 0 |
| | Zinc | 998 | 998 | 998 | 998 | 900 |
| INFRARED | Carbon | 0.41 | 0.14 | 0.33 | 0.18 | 0.11 |
| ABSORBANCE | Water | 0.01 | -0.00 | 0.02 | 0.00 | -0.01 |
| UNITS | Oxidation | 0.37 * | 0.11 | 0.37 * | 0.16 | 0.08 |
| | Nitration | 0.19 | 0.06 | 0.21 * | 0.10 | 0.06 |
| | Blowby | 0.00 | 0.00 | 0.00 | 0.00 | 0.00 |
| | Additive | -0.17 | -0.14 | -0.19 | -0.04 | -0.01 |
| | Nitration | 0.00 | 0.00 | 0.00 | 0.00 | 0.00 |
| | Unsaturate | 0.39 * | 0.12 | 0.34 * | 0.17 | 0.08 |
| | Glycol | 0.00 | 0.01 | 0.00 | 0.00 | 0.01 |
| | Glycol | 0.00 | 0.00 | 0.00 | 0.00 | 0.00 |
| | Additive | -0.06 | -0.05 | -0.04 | -0.02 | 0.00 |
| | Fuel | 0.00 | 0.00 | 0.00 | 0.00 | 0.00 |
| | Fuel | 0.00 | 0.00 | 0.00 | 0.00 | 0.00 |
| VISCOSITY (cSt AT 40C) | | 166.1 | 112.3 | 163.3 | 132.8 | 122.7 |
| (cSt AT 100C) | | 15.0 * | 12.0 | 14.8 * | 13.2 * | 12.6 * |
| RESULT: | | R 32 | W 57 | R 32 | W 57 | W 57 |

CONDITION, 27 APR 89:  C R I T I C A L .

DISCREPANCY: LUBRICANT SUFFERING THERMAL DEGRADATION.
RECOMMENDATION: CHECK OPERATING TEMPERATURE, CHANGE OIL, SERVICE FILTERS.
RESAMPLE: BY 05 MAY (OR IN 87 OPERATING HOURS), TO VERIFY PROBLEM CORRECTED.

**Figure 10.4**  Unit condition report: Front. (Courtesy of Spectron International, Inc.)

CORRECTIVE ACTION INSTRUCTIONS: MONITOR NO. 228        REV: 1

CLIENT: AGRICULTURAL (SUGAR) OPERATION                    DIVISION: TRACTORS
CLIENT ID.: 2050              UNIT DESCRIPTION: CASE TRACTOR    COMPONENT: ENGINE
MODEL: CASE 1896              LUBRICANT: TEXACO URSA SP-30
------------------------------------------------------------------------------------------

CONDITION, 27 APR 89:  C R I T I C A L .    THIS UNIT SHOULD RECEIVE IMMEDIATE ATTENTION

This oil sample shows signs of thermal breakdown. The viscosity is abnormally high, at 137% of normal.
Infrared analysis indicates oxidation/polymerization of the oil, a condition due to excessive temperature.

The operating temperature of the unit should be checked immediately. The oil should be changed and any
filters in the system serviced.

Resample the unit by 05 MAY (or in 87 operating hours), to verify that the problem has been corrected.


PLEASE ADVISE THE LABORATORY OF YOUR FINDINGS.




        *    *    *    *    *    *    *    *    *    *    *


INSPECTION FINDINGS: _____
                     _____
                     _____

CORRECTIVE ACTION TAKEN: _____
                     _____
                     _____


NAME: _____
DATE: _____

**Figure 10.5**   Unit condition report: back. (Courtesy of Spectron International, Inc.)

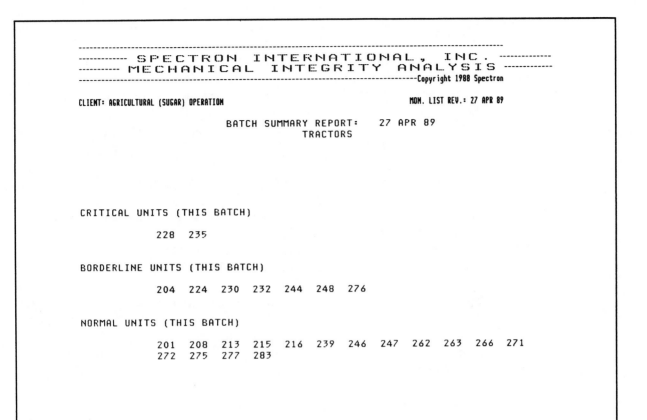

CRITICAL UNITS (THIS BATCH)

        228   235

BORDERLINE UNITS (THIS BATCH)

        204   224   230   232   244   248   276

NORMAL UNITS (THIS BATCH)

        201   208   213   215   216   239   246   247   262   263   266   271
        272   275   277   283

**Figure 10.6**  Batch summary report. (Courtesy of Spectron International, Inc.)

The first page, Fig. 10.6, simply lists the components, by monitor number, in one of the three classifications. Page 2 lists those units classified as critical and lists the information from the bottom of the unit condition report page for each monitor number in this classification. Figure 10.7 illustrates this page. Figure 10.8 lists, in the same manner, those monitored components classified as borderline. In both cases, information is provided that identifies exactly the component in question and the type of lubricant being used. Management thus knows precisely what equipment is being reported on as well as the problem noted and the remedial action recommended in each case. It is obviously not necessary to list normal units.

Figure 10.9 deals with maintenance follow-up on critical units for the month to date. They are grouped by the discrepancies, or problems, causing them to be listed as critical. In the example shown, the causes are silicon, water, oxidation, additive depletion, and incomplete combustion. The manager is thus provided with a clear picture of where the problem areas are and is armed with enough data to enable the manager to ask the right questions of the service personnel. If any units show up chronically with the same problems, they are flagged and the manager has the opportunity to get the situation resolved long before permanent damage occurs to equipment. The problem may be of a technical nature, or it may be a personnel difficulty. The batch summary report is a reliable aid to management. It ensures that things do not get overlooked.

The last pages of this report list those components scheduled for sampling during that particular month and also flags units classified as neglected. Refer to Fig. 10.10. The next page, Fig. 10.11, lists components to be resampled. Items classified as neglected generally denote a problem with personnel.

The unit condition report and the batch summary report form the heart of the mechanical systems integrity program. The former gives detailed information and instructions to the operating service personnel, while the latter provides the needed information to the managers, allowing them to keep abreast of the status of equipment and performance of personnel. Nothing is allowed to slip through the cracks and be overlooked.

The service clerk must ensure that the correct action is carried out on components reported as critical or borderline cases. The clerk should flag the status boards to show those units that require resampling and when this is to be carried out. Copies of unit condition reports should be kept behind the component's index card on the board if they are critical or borderline. Prior reports can be destroyed so as not to accumulate out-of-date reports. Reports that should be retained are identified automatically on the first sheet.

### Monthly Evaluation Report

The monthly evaluation report summarizes the results of the program in two sections. This report is sent to the next-higher level of management, either the general manager or the maintenance manager, depending on the size and organization of the client company. The report is cumulative and automatically replaces prior monthly reports. Some clients keep all monthly

```
-------------------------------------------------------------------------------
----------- S P E C T R O N   I N T E R N A T I O N A L ,   I N C . ------------
---------- M E C H A N I C A L   I N T E G R I T Y   A N A L Y S I S -----------
-----------------------------------------------------------------Copyright 1988 Spectron

CLIENT: AGRICULTURAL (SUGAR) OPERATION                    MON. LIST REV.: 27 APR 89


                    BATCH SUMMARY REPORT:     27 APR 89
                              TRACTORS



CRITICAL UNITS (THIS BATCH)


MONITOR NO.: 235    COLOR: R    CODE:  4
UNIT: 2078   CAT D-5 LGP TRACTOR    ENGINE   CAT-3306    TEXACO URSA SP-30
DISCREPANCY: Coolant chemicals and greater than 0.5% suspended water present in lubricant.
RECOMMENDATION: Inspect for internal leak, change oil, service filters.
RESAMPLE: By 05 MAY (or in 25 operating hours), to verify problem corrected.
ACTION TAKEN:




MONITOR NO.: 228    COLOR: R    CODE: 32
UNIT: 2050   CASE TRACTOR    ENGINE    CASE 1896    TEXACO URSA SP-30
DISCREPANCY: Lubricant suffering thermal degradation.
RECOMMENDATION: Check operating temperature, change oil, service filters.
RESAMPLE: By 05 MAY (or in 87 operating hours), to verify problem corrected.
ACTION TAKEN:
```

**Figure 10.7**   Batch summary report. (Courtesy of Spectron International, Inc.)

```
--------------------------------------------------------------------------------
------------- S P E C T R O N   I N T E R N A T I O N A L ,   I N C . -------------
----------- M E C H A N I C A L   I N T E G R I T Y   A N A L Y S I S -------------
--------------------------------------------------------------Copyright 1988 Spectron

CLIENT: AGRICULTURAL (SUGAR) OPERATION                    MON. LIST REV.: 27 APR 89

                          BATCH SUMMARY REPORT:    27 APR 89
                                  TRACTORS


        BORDERLINE UNITS (THIS BATCH)


        MONITOR NO.: 248    COLOR: Y    CODE: 27
        UNIT: 2111    J DEERE TRACTOR 4050    ENGINE    JD RG 6466D    TEXACO URSA SP-30
        DISCREPANCY: Abnormal wear levels indicated.
        RECOMMENDATION: Monitor at reduced interval.
        RESAMPLE: By 12 MAY (or in 175 operating hours), to maintain close surveillance.
        ACTION TAKEN:



        MONITOR NO.: 276    COLOR: Y    CODE: 27
        UNIT: 2184    CASE 9150    ENGINE    CUMMINS LTA10    TEXACO URSA SP-30
        DISCREPANCY: Abnormal wear levels indicated.
        RECOMMENDATION: Monitor at reduced interval.
        RESAMPLE: By 12 MAY (or in 50 operating hours), to maintain close surveillance.
        ACTION TAKEN:



        MONITOR NO.: 230    COLOR: Y    CODE: 29
        UNIT: 2052    CASE TRACTOR    ENGINE    CASE 1896    TEXACO URSA SP-30
        DISCREPANCY: Abnormal lubricant condition indicated.
        RECOMMENDATION: Monitor at reduced interval.
        RESAMPLE: By 05 MAY (or in 87 operating hours), to maintain close surveillance.
        ACTION TAKEN:



        MONITOR NO.: 232    COLOR: Y    CODE: 29
        UNIT: 2054    CASE TRACTOR    ENGINE    CASE 1896    TEXACO URSA SP-30
        DISCREPANCY: Abnormal lubricant condition indicated.
        RECOMMENDATION: Monitor at reduced interval.
        RESAMPLE: By 05 MAY (or in 87 operating hours), to maintain close surveillance.
        ACTION TAKEN:



        MONITOR NO.: 224    COLOR: Y    CODE: 37
        UNIT: 2043    J DEERE TRACTOR 4440    ENGINE    JD 6466 TR    TEXACO URSA SP-30
        DISCREPANCY: Lubricant antiwear additives excessively depleted.
        RECOMMENDATION: Change oil and service filters.
        RESAMPLE: By 05 MAY (or in 87 operating hours), to verify problem corrected.
        ACTION TAKEN:
```

**Figure 10.8**  Batch summary report. (Courtesy of Spectron International, Inc.)

```
-----------------------------------------------------------------------
------------ S P E C T R O N   I N T E R N A T I O N A L ,   I N C . -------------
---------- M E C H A N I C A L   I N T E G R I T Y   A N A L Y S I S -------------
-------------------------------------------------------------Copyright 1988 Spectron

CLIENT: AGRICULTURAL (SUGAR) OPERATION                    MON. LIST REV.: 27 APR 89

                    BATCH  SUMMARY  REPORT:    27 APR 89
                              TRACTORS


MAINTENANCE FOLLOWUP: CRITICAL UNITS


At this time the following units are classified as CRITICAL.  They should receive immediate maintenance attention
if any is specified, and should be resampled within two weeks of operation or less.  An asterisk (*) indicates
that a unit's problem is chronic.  Units are grouped by problem type.



CRITICAL UNITS: WATER

MON. NO.             UNIT              SINCE      MON. NO.              UNIT               SINCE
--------  -----------------------------  ---------  --------  -----------------------------  ---------

  235    2078  CAT D-5 LGP TRACTOR   ENGINE  27 Apr 89


CRITICAL UNITS: SILICON

MON. NO.             UNIT              SINCE      MON. NO.              UNIT               SINCE
--------  -----------------------------  ---------  --------  -----------------------------  ---------

  268    2176  STEIGER 9150C TRAC   ENGINE  14 Apr 89


CRITICAL UNITS:  INCOMPLETE COMBUSTION

MON. NO.             UNIT              SINCE      MON. NO.              UNIT               SINCE
--------  -----------------------------  ---------  --------  -----------------------------  ---------

  214    2029  J DEERE TRACTOR 4050 ENGINE  14 Apr 89    217    2032  J DEERE TRACTOR 4050  ENGINE  14 Apr 89



CRITICAL UNITS:  OXIDATION

MON. NO.             UNIT              SINCE      MON. NO.              UNIT               SINCE
--------  -----------------------------  ---------  --------  -----------------------------  ---------

  212    2027  J DEERE TRACTOR 4050 ENGINE  14 Apr 89    228    2050  CASE TRACTOR         ENGINE  27 Apr 89
  229    2051  CASE TRACTOR         ENGINE  14 Apr 89    231 *  2053  CASE TRACTOR         ENGINE  14 Apr 89
  281    2058  CASE TRACTOR         ENGINE  14 Apr 89    284    2061  CASE TRACTOR         ENGINE  14 Apr 89



CRITICAL UNITS:  ADDITIVE DEPLETION

MON. NO.             UNIT              SINCE      MON. NO.              UNIT               SINCE
--------  -----------------------------  ---------  --------  -----------------------------  ---------

  203    2015  J DEERE TRACTOR 4040 ENGINE  14 Apr 89
```

**Figure 10.9**  Batch summary report. (Courtesy of Spectron International, Inc.)

```
----------------------------------------------------------------------------
------------ S P E C T R O N   I N T E R N A T I O N A L ,   I N C .  ------------
---------- M E C H A N I C A L   I N T E G R I T Y   A N A L Y S I S  ------------
--------------------------------------------------------------------Copyright 1988 Spectron

CLIENT: AGRICULTURAL (SUGAR) OPERATION                    MON. LIST REV.: 27 APR 89

                    BATCH  SUMMARY  REPORT:    27  APR  89
                              TRACTORS


       MAINTENANCE FOLLOWUP: UNSAMPLED UNITS (MONTH TO DATE)


       The following units are now scheduled for sampling.  An asterisk (*) indicates that a unit is classified as
       neglected (three or more months overdue for sampling).



       MON. NO.               UNIT            MON. NO.              UNIT
       --------  ---------------------------------   --------  ---------------------------------
         274    2182             CASE 9150C
```

**Figure 10.10**   Batch summary report. (Courtesy of Spectron International, Inc.)

```
-------------------------------------------------------------------------
----------- S P E C T R O N   I N T E R N A T I O N A L ,   I N C . ------------
---------- M E C H A N I C A L   I N T E G R I T Y   A N A L Y S I S ------------
-------------------------------------------------------------Copyright 1988 Spectron

CLIENT: AGRICULTURAL (SUGAR) OPERATION                    MON. LIST REV.: 27 APR 89

                    B A T C H   S U M M A R Y   R E P O R T :    27 APR 89
                                   TRACTORS


MAINTENANCE FOLLOWUP: UNITS TO BE RESAMPLED


The following units are currently classified as CRITICAL or BORDERLINE and should be resampled no later than
the date labeled as DUE BY.  If the DUE BY date for a unit has expired then the unit must be resampled
immediately.


CRITICAL UNITS


MON. NO.            UNIT            DUE BY      MON. NO.               UNIT              DUE BY
--------   ---------------------   --------    --------   ----------------------------  ---------
  206    2019  J DEERE TRACTOR 4040  ENGINE   IMMED        242    2104  J DEERE TRACTOR 4040  ENGINE  22 Apr 89
  203    2015  J DEERE TRACTOR 4040  ENGINE  22 Apr 89     268    2176  STEIGER 9150C TRAC    ENGINE  22 Apr 89
  212    2027  J DEERE TRACTOR 4050  ENGINE  22 Apr 89     280    2057  CASE TRACTOR          ENGINE  22 Apr 89
  214    2029  J DEERE TRACTOR 4050  ENGINE  22 Apr 89     281    2058  CASE TRACTOR          ENGINE  22 Apr 89
  217    2032  J DEERE TRACTOR 4050  ENGINE  22 Apr 89     284    2061  CASE TRACTOR          ENGINE  22 Apr 89
  229    2051  CASE TRACTOR          ENGINE  22 Apr 89     228    2050  CASE TRACTOR          ENGINE   5 May 89
  231    2053  CASE TRACTOR          ENGINE  22 Apr 89     235    2078  CAT D-5 LGP TRACTOR   ENGINE   5 May 89
  240    2102  J DEERE TRACTOR 4040  ENGINE  22 Apr 89


BORDERLINE UNITS


MON. NO.            UNIT            DUE BY      MON. NO.               UNIT              DUE BY
--------   ---------------------   --------    --------   ----------------------------  ---------
  209    2024  J DEERE TRACTOR 4050  ENGINE  22 Apr 89     223    2042  J DEERE TRACTOR 4440  ENGINE  29 Apr 89
  211    2026  J DEERE TRACTOR 4050  ENGINE  22 Apr 89     253    2123  J DEERE TRACTOR 4440  ENGINE  29 Apr 89
  222    2041  J DEERE TRACTOR 4440  ENGINE  22 Apr 89     224    2043  J DEERE TRACTOR 4440  ENGINE   5 May 89
  249    2112  J DEERE TRACTOR 4050  ENGINE  22 Apr 89     230    2052  CASE TRACTOR          ENGINE   5 May 89
  250    2113  J DEERE TRACTOR 4050  ENGINE  22 Apr 89     232    2054  CASE TRACTOR          ENGINE   5 May 89
  251    2114  J DEERE TRACTOR 4050  ENGINE  22 Apr 89     244    2106  J DEERE TRACTOR 4040  ENGINE   5 May 89
  252    2120  J DEERE TRACTOR 4250  ENGINE  22 Apr 89     248    2111  J DEERE TRACTOR 4050  ENGINE  12 May 89
  261    2150  CAT D5 LGPC TRACTOR   ENGINE  22 Apr 89     276    2184  CASE 9150             ENGINE  12 May 89
  286    2063  CASE TRACTOR          ENGINE  22 Apr 89     204    2017  J DEERE TRACTOR 4040  ENGINE  27 May 89
  207    2022  J DEERE TRACTOR 4040  ENGINE  29 Apr 89
```

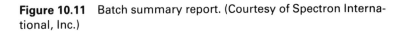

**Figure 10.11**  Batch summary report. (Courtesy of Spectron International, Inc.)

evaluation reports through the December, or twelfth, report, then discard all but this report, which is, in effect, the year's summary.

*Section I.* Section I presents management with an overview of the program's results with data presented in graphic form. It shows results for the month, year to date, and the prior year. Thus benchmarks are provided against which management can measure the effectiveness of the maintenance operation in terms of mechanical condition of equipment and supervisory control. Performance is measured using key indicators plotted graphically to reveal trends as the year progresses. Figure 10.12 lists those indicators of performance.

Figures 10.13 through 10.15 graphically illustrate each indicator of mechanical condition by month for both the present year, shown in shaded columns, and the preceding year, shown as open columns. Figures 10.16 through 10.18 provide similar information for those indicators of supervisory control over the program.

The last page of Section I of the monthly evaluation report rates the operation for the month in three categories. There are four ratings in each category. Shaded columns represent the current year and open columns the preceding year. Figure 10.19 illustrates a typical rating page.

*Section II.* Section II of the monthly evaluation report consists of detailed tabulations showing component status, changes in condition, units requiring special attention, and critical component discrepancies. This information is first presented in aggregate form to allow interdivisional comparisons, then represented by division. The divisional reports list individual components by monitor number. Those companies whose operations are small and not broken into separate divisions will receive one division report. The single division report is not simply a rehash of the aggregate: It identifies, by monitor number, individual components requiring surveillance. This section of the report provides the following information under the aggregate heading.

*A. Monitored Equipment Status Summary:* This provides a summation of the monitored components for the month, grouping them into categories to reflect the overall equipment condition. This is illustrated in Fig. 10.20.

*B. Units Requiring Special Attention:* This indicates components requiring special attention at the supervisory level and is illustrated in Fig. 10.21.

*C. Changes in Unit Condition:* Figure 10.22 illustrates this page, indicating units that have either improved or deteriorated since their previous monitoring.

*D. Units Not Requiring Sampling During Current Month:* Figure 10.22 also lists the number of components not scheduled for monitoring and the reasons they were excluded.

*E. Critical Units Summary:* This is shown in Fig. 10.23 and is meant as a guide to focus maintenance priorities and resources.

# spectron international, inc.

AGRICULTURAL (SUGAR) OPERATION                    November 1988    p.   2

MONTHLY EVALUATION REPORT - PART I.

      This section reviews performance of the maintenance operation for the month and year-to-date in terms of MECHANICAL CONDITION of monitored systems and the SUPERVISORY CONTROL of the program.  Performance is measured using key indicators which are plotted graphically to provide a concise overview and to reveal trends as the year progresses.  These key indicators are as follows:

| MECHANICAL CONDITION | SUPERVISORY CONTROL |
|---|---|
| Availability | Sampling Coverage |
| Critical Units | Neglected Units |
| Borderline Units | Chronic Units |

**Figure 10.12** Monthly evaluation report: part I. (Courtesy of Spectron International, Inc.)

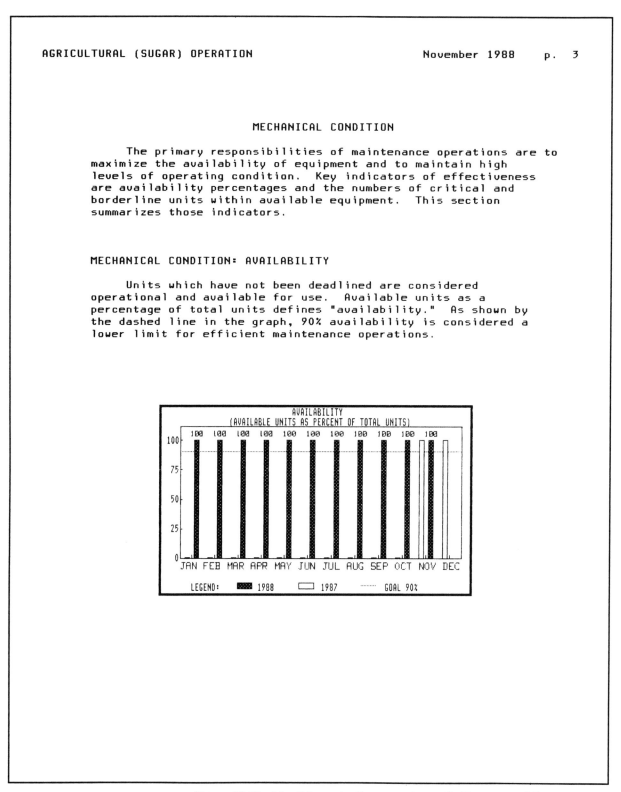

AGRICULTURAL (SUGAR) OPERATION                    November 1988    p.  3


MECHANICAL CONDITION

        The primary responsibilities of maintenance operations are to
maximize the availability of equipment and to maintain high
levels of operating condition.  Key indicators of effectiveness
are availability percentages and the numbers of critical and
borderline units within available equipment.  This section
summarizes those indicators.


MECHANICAL CONDITION: AVAILABILITY

        Units which have not been deadlined are considered
operational and available for use.  Available units as a
percentage of total units defines "availability."  As shown by
the dashed line in the graph, 90% availability is considered a
lower limit for efficient maintenance operations.

**Figure 10.13** Monthly evaluation report: part I. (Courtesy of Spectron International, Inc.)

AGRICULTURAL (SUGAR) OPERATION                    November 1988    p.   4


MECHANICAL CONDITION: CRITICAL UNITS

        Units whose samples indicate radical departures from normal
condition are classified as critical.  These units require
priority corrective action (as specified in UNIT CONDITION
REPORTS) to prevent further deterioration.  Prompt attention is
usually sufficient to restore such units to acceptable mechanical
condition without major expenditure of maintenance effort.  The
dashed line at 5% in the graph indicates what is considered the
upper limit for critical units consistent with effective
maintenance.

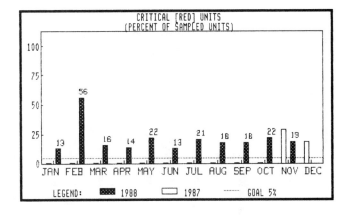

**Figure 10.14**  Monthly evaluation report: part I. (Courtesy of Spectron International, Inc.)

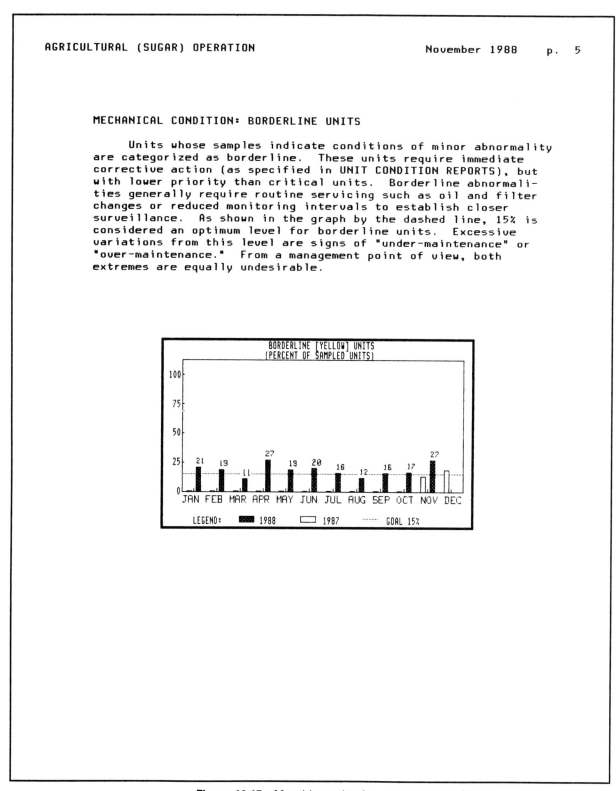

**Figure 10.15** Monthly evaluation report: part I. (Courtesy of Spectron International, Inc.)

AGRICULTURAL (SUGAR) OPERATION                    November 1988    p.   6

### SUPERVISORY CONTROL

        Overall, the level of MECHANICAL CONDITION is directly
related to the level of SUPERVISORY CONTROL provided within the
maintenance operation. Key indicators of control are the monthly
sampling coverage and the numbers of neglected and chronic units
each month.  This sub-section summarizes these key indicators.


SUPERVISORY CONTROL: SAMPLING COVERAGE

        Sampling coverage is of prime importance to the maintenance
program.  It provides timely knowledge of actual mechanical and
lubricant condition of each unit, which is required for effective
maintenance management.  As shown in the graph, 100% is the
optimum level of sampling coverage.

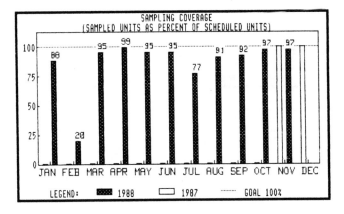

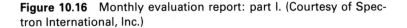

**Figure 10.16**  Monthly evaluation report: part I. (Courtesy of Spectron International, Inc.)

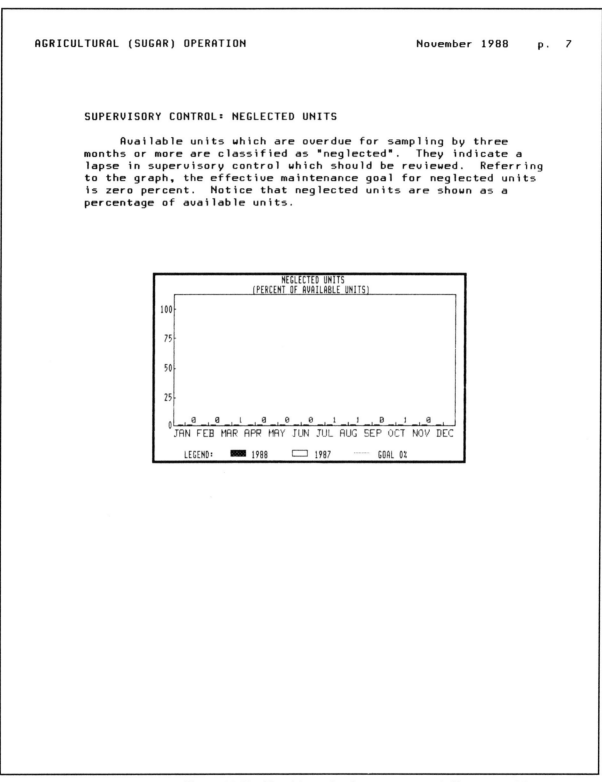

AGRICULTURAL (SUGAR) OPERATION                    November 1988    p.  7


   SUPERVISORY CONTROL: NEGLECTED UNITS

      Available units which are overdue for sampling by three
   months or more are classified as "neglected".  They indicate a
   lapse in supervisory control which should be reviewed.  Referring
   to the graph, the effective maintenance goal for neglected units
   is zero percent.  Notice that neglected units are shown as a
   percentage of available units.

NEGLECTED UNITS
(PERCENT OF AVAILABLE UNITS)

100

 75

 50

 25

  0
   JAN FEB MAR APR MAY JUN JUL AUG SEP OCT NOV DEC

   LEGEND:    1988      1987      GOAL 0%

**Figure 10.17**  Monthly evaluation report: part I. (Courtesy of Spectron International, Inc.)

AGRICULTURAL (SUGAR) OPERATION          November 1988    p.    8

SUPERVISORY CONTROL: CHRONIC UNITS

     Of the units which have been reported as critical, those
which have received at least three  consecutive  critical reports
are reclassified as "chronic".  Such units should be removed from
service immediately to determine why the critical condition
persists.  This graph shows chronic units as a percentage of
available units, with zero percent as the effective maintenance
goal.

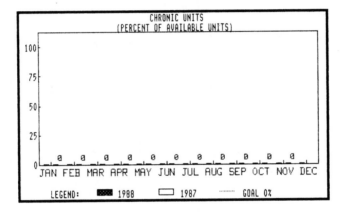

**Figure 10.18**  Monthly evaluation report: part I. (Courtesy of Spectron International, Inc.)

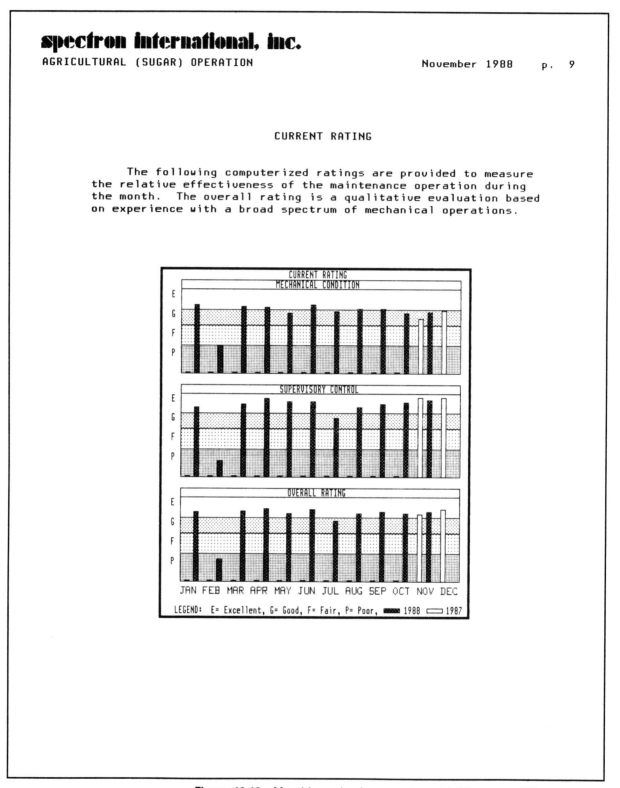

**Figure 10.19**  Monthly evaluation report: part I. (Courtesy of Spectron International, Inc.)

AGRICULTURAL (SUGAR) OPERATION                    November 1988    p. 12

A.  MONITORED EQUIPMENT STATUS SUMMARY
    The equipment being monitored is divided into two main groups,
    those available for operation and those deadlined for repairs.
    The available units are further divided for reporting purposes
    into two subgroups, either scheduled for sampling or unscheduled.
    Of the units scheduled for sampling, some are reported by the
    client as being on standby, thus, not requiring sampling during
    the current month.  Of the remainder, sampled units are
    classified as normal, borderline or critical, or, possibly, being
    emphasized as chronic.  Units scheduled for sampling but not
    sampled are classified as unsampled, unless they are overdue for
    sampling by at least three months, which places them into the
    neglected classification.  Deadlined units are as reported by
    the client.

### MONITORED EQUIPMENT STATUS SUMMARY

| CATEGORY | TOTAL | PERCENTAGES |
|---|---|---|
| TOTAL MONITORED UNITS | 148 | 100% OF TOTAL |
| DEADLINED UNITS | 0 | 0% OF TOTAL |
| AVAILABLE UNITS | 148 | 100% OF TOTAL |
| Scheduled Units | 93 | 63% of available |
| Sampled Units | 90 | 97% of scheduled |
| Normal (White) | 49 | 54% of sampled |
| Borderline(Yellow) | 24 | 27% of sampled |
| CRITICAL (Red) | 17 | 19% of sampled |
| CHRONIC *** | 0 | 0% of available |
| Unsampled Units | 3 | 3% of scheduled |
| NEGLECTED *** | 0 | 0% of available |
| Unscheduled Units | 49 | 33% of available |
| Standby Units | 3 | 2% of available |
| Sched. Shutdown Units | 3 | 2% of available |

**Figure 10.20**  Monthly evaluation report: part II. (Courtesy of Spectron International, Inc.)

```
AGRICULTURAL (SUGAR) OPERATION                    November 1988    p. 13

   B.   UNITS REQUIRING SPECIAL ATTENTION

      The following units require special attention from supervisory
      personnel.  CHRONIC units are those having received 3 consecutive
      critical (red) reports.  They require IMMEDIATE maintenance
      follow-up to prevent equipment failure.  NEGLECTED units are
      those which are three (3) or more months overdue for sampling.
      DEADLINED units and STANDBY units must be so reported by the
      client or they will become classified as UNSAMPLED or NEGLECTED
      units.
```

| | | D | W | T | N |
| | | I | A | R | E |
| | | V | T | A | W |
| | | I | E | C | |
| | | S | R | T | O |
| | | I | | O | I |
| | | O | C | R | L |
| | | N | O | S | |
| | | S | N | | S |
| | | | T | | T |
| | | | R | | O |
| | | | O | | R |
| | | | L | | A |
| | | | | | G |
| | | | | | E |
| CATEGORY | TOTAL | | | | |
| CHRONIC | 0 | 0 | 0 | 0 | |
| NEGLECTED | 0 | 0 | 0 | 0 | |
| DEADLINED | 0 | 0 | 0 | 0 | |
| OTHER UNSAMPLED | 3 | 1 | 1 | 1 | |

**Figure 10.21**   Monthly evaluation report: part II. (Courtesy of Spectron International, Inc.)

AGRICULTURAL (SUGAR) OPERATION                    November 1988    p. 14

   C.  CHANGES IN UNIT CONDITION

      Trends in maintenance effectiveness can be determined from a
      review of improvements and deteriorations in unit condition.  If
      the number of deteriorated units exceeds the number of improved
      units, an adverse trend may be developing.  Improvement is
      defined as the change in a unit's condition from critical to
      borderline or normal, while the transition from normal or
      borderline to critical is defined as a deterioration.

| CATEGORY | TOTAL | DIVISIONS | WATER CONTROL | TRACTORS | NEW OIL STORAGE |
|---|---|---|---|---|---|
| Deteriorated | 9 | 4 | 5 | 0 | |
| Improved | 3 | 1 | 2 | 0 | |

   D.  UNITS NOT REQUIRING SAMPLING DURING CURRENT MONTH

| CATEGORY | TOTAL | | | |
|---|---|---|---|---|
| Standby | 3 | 0 | 3 | 0 |
| Unscheduled | 49 | 42 | 7 | 0 |
| Sched. Shutdown | 3 | 3 | 0 | 0 |

**Figure 10.22**  Monthly evaluation report: part II. (Courtesy of Spectron International, Inc.)

AGRICULTURAL (SUGAR) OPERATION                    November 1988    p. 15

    E.   CRITICAL UNITS SUMMARY
         The most prevalent problems during the month can be determined
         from this listing, and should be given particular and immediate
         attention by the maintenance group.

| DISCREPANCY | TOTAL | DIVISIONS | WATER CONTROL | TRACTORS | NEW OIL STORAGE |
|---|---|---|---|---|---|
| Water | 0 | 0 | 0 | 0 | |
| Silicon | 1 | 0 | 1 | 0 | |
| Silicon & Fuel Dilution | 1 | 1 | 0 | 0 | |
| Fuel Dilution | 0 | 0 | 0 | 0 | |
| Incomplete Combustion | 7 | 3 | 4 | 0 | |
| Bearing Wear | 0 | 0 | 0 | 0 | |
| Oxidation | 3 | 0 | 3 | 0 | |
| Oxidation/Nitration | 0 | 0 | 0 | 0 | |
| Additive Depletion | 3 | 0 | 3 | 0 | |
| Entrained Gases | 0 | 0 | 0 | 0 | |
| Improper Lubricant | 0 | 0 | 0 | 0 | |
| Blowby | 0 | 0 | 0 | 0 | |
| Wear Surveillance | 2 | 0 | 2 | 0 | |
| Lubricant Surveillance | 0 | 0 | 0 | 0 | |
| Missing Information | 0 | 0 | 0 | 0 | |

**Figure 10.23**  Monthly evaluation report: part II. (Courtesy of Spectron International, Inc.)

It should be noted that in the aggregate portion of the report, only the quantity of units fitting a particular category are listed.

In the divisional section, the units are identified individually by their monitor number. Figures 10.24 through 10.26 are examples of the information privided in this section. They cover the following topics:

*F. Critical Unit Discrepancies:*   See Fig. 10.24.

*G. Units Requiring Special Attention:*   See Fig. 10.25.

*H. Units with Deteriorated or Improved Condition:*   See Fig. 10.26.

*I. Units Not Requiring Sampling During Current Month:*   See Fig. 10.26.

Items F through I should be used by the manager to ensure that the maintenance staff is taking action to correct the indicated problems.

These routine reports provide the needed information to all levels of management and the work force to ensure that the equipment is kept operating at peak efficiency. In summation, higher management requires both frequent and long-term audit reports of all aspects of the operation, indicating effectiveness, focusing attention on trends, and identifying problem areas. The service manager and staff need more detailed information about the day-to-day status of equipment and the maintenance program. Shop and field personnel must have very detailed information, on a daily basis, as to which specific units need attention and an exact description of the corrective steps to be taken in each case.

As these reports are studied and put to use by the various people who receive them, it will not take long to identify where the biggest problems are and begin action to eliminate them. Information and ideas will start to flow between management, service, and lab staffs. A spirit of cooperation in dealing with machinery and maintenance procedures will develop, primarily because the mystery has been removed from the inner workings of that machinery. The status of each unit is known and understood. If normal corrective procedures should not rectify a particular chronic problem, the consulting lab will be ready to work with the client company to arrive at a solution. This is where special studies and reports become most useful.

## Special Reports

There are two types of reports that are normally designed to cover a year of operation. This time period is used because it coincides with normal business cycles, and it allows sufficient data to be accumulated to give accurate results. The information generated by a program of mechanical systems integrity management serves as a basis for a wide variety of special studies and comparative evaluations. The two most commonly used are statistical wear profiles and confidence factor analysis.

*Statistical Wear Profiles.*   These tables provide a convenient means of determining wear profiles of equipment monitored during a prolonged period. The profiles enable each program to be customized for the individual client. They also serve as the basis for comparative evaluations of equipment, lubricants, filters, mechanical components, and maintenance and

```
AGRICULTURAL (SUGAR) OPERATION                    November 1988    p. 16-2


       F.  CRITICAL UNIT DISCREPANCIES
           The most prevalent problems during the month can be determined
           from this listing, and should be given particular and immediate
           attention by the maintenance group.


                                  WATER CONTROL
```

| DISCREPANCY | TOTAL | UNIT NO.S |
|---|---|---|
| Water | 0 | |
| Silicon | 0 | |
| Silicon & Fuel Dilution | 1 | 66 |
| Fuel Dilution | 0 | |
| Incomplete Combustion | 3 | 15  48  49 |
| Bearing Wear | 0 | |
| Oxidation | 0 | |
| Oxidation/Nitration | 0 | |
| Additive Depletion | 0 | |
| Entrained Gases | 0 | |
| Improper Lubricant | 0 | |
| Blowby | 0 | |
| Wear Surveillance | 0 | |
| Lubricant Surveillance | 0 | |
| Missing Information | 0 | |

**Figure 10.24** Monthly evaluation report: part II. (Courtesy of Spectron International, Inc.)

```
AGRICULTURAL (SUGAR) OPERATION                    November 1988    p. 16-3

    G.  UNITS REQUIRING SPECIAL ATTENTION

        The following units require special attention from supervisory
        personnel.  CHRONIC units are those having received 3 consecutive
        critical (red) reports.  They require IMMEDIATE maintenance
        follow-up to prevent equipment failure.  NEGLECTED units are
        those which are three (3) or more months overdue for sampling.
        DEADLINED units and STANDBY units must be so reported by the
        client or they will become classified as UNSAMPLED or NEGLECTED
        units.

                              WATER CONTROL
        _____
            CATEGORY        TOTAL  |                UNIT NO.S
                                   |
        CHRONIC               0    |
                                   |
                                   |
        NEGLECTED             0    |
                                   |
                                   |
        DEADLINED             0    |
                                   |
                                   |
        OTHER UNSAMPLED       1    | 77
                                   |
                                   |
```

**Figure 10.25** Monthly evaluation report: part II. (Courtesy of Spectron International, Inc.)

```
AGRICULTURAL (SUGAR) OPERATION                        November 1988    p. 16-4

    H.   UNITS WITH DETERIORATED OR IMPROVED CONDITION

        Trends in maintenance effectiveness can be determined from a
        review of improvements and deteriorations in unit condition.  If
        the number of deteriorated units exceeds the number of improved
        units, an adverse trend may be developing.  Improvement is
        defined as the change in a unit's condition from critical to
        borderline or normal, while the transition from normal or
        borderline to critical is defined as a deterioration.


                              WATER CONTROL
```

| CATEGORY | TOTAL | UNIT NO.S |
|----------|-------|-----------|
| Deteriorated | 4 | 15  48  49  66 |
| Improved | 1 | 62 |

```
    I.   UNITS NOT REQUIRING SAMPLING DURING CURRENT MONTH
```

| CATEGORY | TOTAL | UNIT NO.S |
|----------|-------|-----------|
| Standby | 0 | |
| Unscheduled | 42 | 5   6   7   9   10  11  13  16  21  25  30  33  34  35 |
| | | 36  38  44  45  46  47  50  51  52  53  55  57  58  59 |
| | | 60  65  68  71  72  73  74  76  81  82  83  84  85  88 |
| Sched. Shutdown | 3 | 56  61  69 |

**Figure 10.26**  Monthly evaluation report: part II. (Courtesy of Spectron International, Inc.)

operational procedures. They are presented as tables and also in graphic form as frequency histograms. Figure 10.27 illustrates a wear profile in table form and Fig. 10.28 shows an example of the histograms. There is a histogram printed out for each trace element. The example shows only the first two for the purpose of illustration.

*Confidence Factor Analysis.* This study ranks equipment according to the level of confidence that can be placed in each component's mechanical integrity based on a review of the data accumulated during the referenced period. The purpose of the ranking is to provide the client company with an insight into the differences in long-term operational reliability that exist in its equipment. The confidence factor is derived from the ratio of critical discrepancy reports to total reports, for a given component, over an extended period.

This analysis is presented in both bar chart and tabular form. The bar chart shows the distribution of components in the different confidence categories. Ideally, all components should appear within the 91 to 100 percent category. A typical bar chart is shown in Fig. 10.29. The table provides a detailed breakdown, identifying each component by monitor number, description, what category it falls into, and the exact confidence factor assigned to it. Figure 10.30 illustrates the first page of a typical table. Categories go from zero up through the highest, designated as the 91 to 100 percent group.

In addition, a summary is included that covers information as to the types of discrepancies detected in the equipment during the period covered by the report. This quickly focuses management's attention on the company's problem areas. Figure 10.31 shows an example of such a summary. This analysis can also be done by equipment type and/or division, in addition to the global or total company analysis.

A second appendix is often included that details the individual sample results accumulated by each component having a confidence factor of 70 percent or less during the reporting period. Figure 10.32 illustrates the first page of a typical report.

Armed with the information contained in the monthly and special reports, management can take action to eliminate costly problems and bring their equipment up to optimum efficiency. If additional technical assistance is needed, the consulting lab can conduct further studies and evaluations to solve specific problems associated with a component, unit, or group of units.

A case involving such assistance involved a small fleet of diesel engine pickup trucks used by the supervisors and maintenance personnel on a large corporate farm. The working environment was extremely dusty. The trucks were new and the unit condition reports were calling for an oil and filter change every two weeks, based on the presence of silicon and accompanying wear metals in the lubricating oil. Servicing the air intake filters and systems provided no improvement. By the end of the fourth month, the consulting lab made a grim prediction based on accumulated data. Should the trend continue unchecked, a major engine overhaul would be needed on each truck by the end of the first season. The lab and company both agreed that the standard air filtering system was inadequate and an extra-heavy-

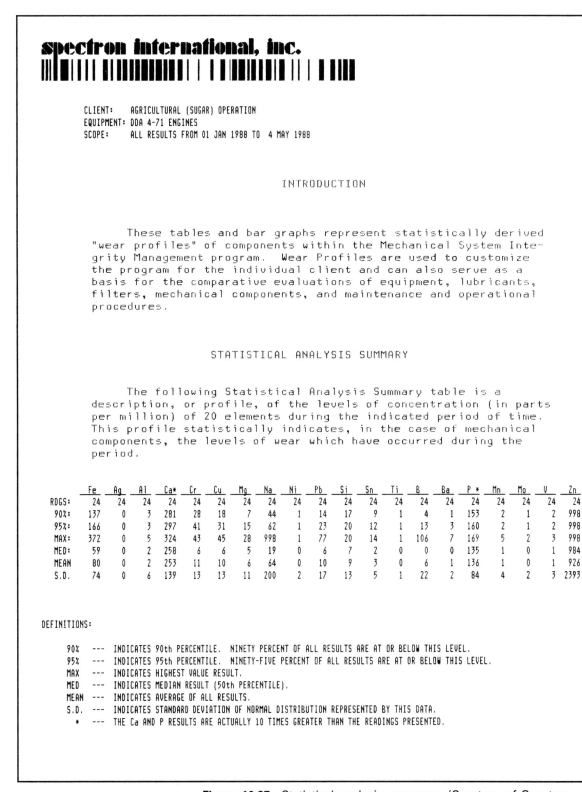

**Figure 10.27**   Statistical analysis summary. (Courtesy of Spectron International, Inc.)

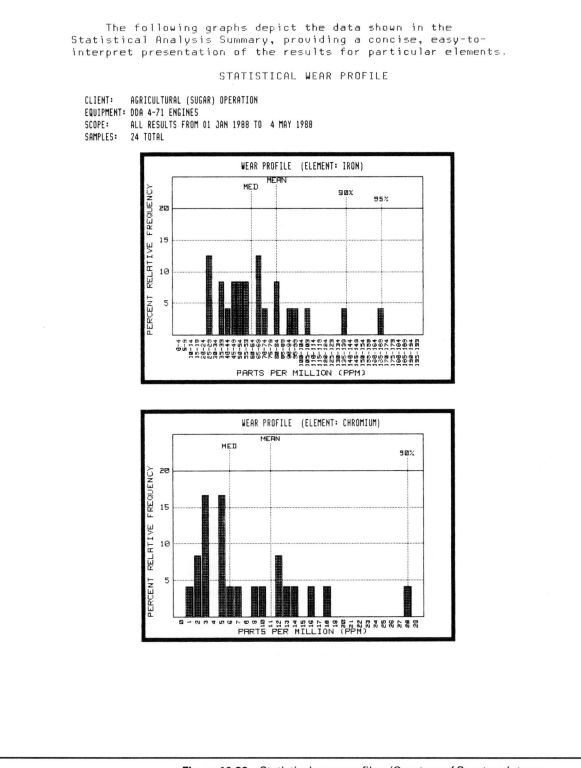

**Figure 10.28** Statistical wear profiles. (Courtesy of Spectron International, Inc.)

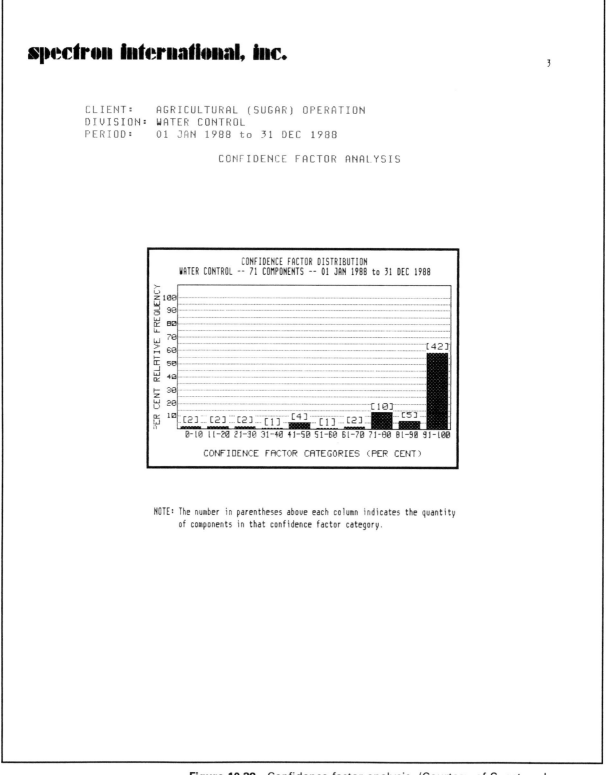

**Figure 10.29**  Confidence factor analysis. (Courtesy of Spectron International, Inc.)

# spectron international, inc.

4

```
            CLIENT:    AGRICULTURAL (SUGAR) OPERATION
            DIVISION:  WATER CONTROL
            PERIOD:    01 JAN 1988 to 31 DEC 1988

                       CONFIDENCE FACTOR ANALYSIS
```

| CATEGORY | MON | R/S RATIO | CONF.FACTOR | COMPONENT IDENTIFICATION | | | |
|----------|-----|-----------|-------------|------|------|------|------|
| 0 - 10 | 68 | 3/3 | 0 | 5068 | WATER PUMP | ENGINE CAT D-330 | TEXACO URSA SP-30 |
| | 15 | 4/4 | 0 | 5015 | WATER PUMP | ENGINE DDA 4-71 | TEXACO URSA SP-30 |
| 11 - 20 | 49 | 4/5 | 20 | 5049 | WATER PUMP | ENGINE CAT D-330 | TEXACO URSA SP-30 |
| | 48 | 4/5 | 20 | 5048 | WATER PUMP | ENGINE CAT D-330 | TEXACO URSA SP-30 |
| 21 - 30 | 69 | 3/4 | 25 | 5069 | WATER PUMP | ENGINE CAT D-330 | TEXACO URSA SP-30 |
| | 57 | 5/7 | 29 | 5057 | WATER PUMP | ENGINE DDA 4-71 | TEXACO URSA SP-30 |
| 31 - 40 | 25 | 3/5 | 40 | 5025 | WATER PUMP | ENGINE DDA 4-71 | TEXACO URSA SP-30 |
| 41 - 50 | 88 | 1/2 | 50 | 5088 | WATER PUMP | ENGINE DDA 6-71 | TEXACO URSA SP-30 |
| | 81 | 4/8 | 50 | 5081 | WATER PUMP | ENGINE DDA 6-71 | TEXACO URSA SP-30 |
| | 66 | 2/4 | 50 | 5066 | WATER PUMP | ENGINE DDA 4-71 | TEXACO URSA SP-30 |
| | 10 | 3/6 | 50 | 5010 | WATER PUMP | ENGINE DDA 3-71 | TEXACO URSA SP-30 |
| 51 - 60 | 9 | 2/5 | 60 | 5009 | WATER PUMP | ENGINE DEUTZ AF4L912 | TEXACO URSA SP-30 |
| 61 - 70 | 30 | 2/6 | 67 | 5030 | WATER PUMP | ENGINE DDA 4-71 | TEXACO URSA SP-30 |
| | 16 | 1/3 | 67 | 5016 | WATER PUMP | ENGINE DDA 4-53 | TEXACO URSA SP-30 |
| 71 - 80 | 7 | 2/7 | 71 | 5007 | WATER PUMP | ENGINE DDA 3-71 | TEXACO URSA SP-30 |
| | 79 | 2/8 | 75 | 5079 | WATER PUMP | ENGINE DDA 6-71 | TEXACO URSA SP-30 |
| | 40 | 1/4 | 75 | 5040 | WATER PUMP | ENGINE CAT D-330 | TEXACO URSA SP-30 |
| | 87 | 1/5 | 80 | 5087 | WATER PUMP | ENGINE DDA 4-71 | TEXACO URSA SP-30 |
| | 76 | 1/5 | 80 | 5076 | WATER PUMP | ENGINE CAT D-333 | TEXACO URSA SP-30 |
| | 74 | 1/5 | 80 | 5074 | WATER PUMP | ENGINE AC 10000 | TEXACO URSA SP-30 |
| | 60 | 1/5 | 80 | 5060 | WATER PUMP | ENGINE DEUTZ F3L912 | TEXACO URSA SP-30 |
| | 44 | 1/5 | 80 | 5044 | WATER PUMP | ENGINE JD 6466DF | TEXACO URSA SP-30 |
| | 36 | 1/5 | 80 | 5036 | WATER PUMP | ENGINE CAT D-330 | TEXACO URSA SP-30 |
| | 34 | 1/5 | 80 | 5034 | WATER PUMP | ENGINE CAT D-315 | TEXACO URSA SP-30 |
| 81 - 90 | 65 | 1/6 | 83 | 5065 | WATER PUMP | ENGINE DEUTZ F3L912 | TEXACO URSA SP-30 |
| | 62 | 1/6 | 83 | 5062 | WATER PUMP | ENGINE DDA 4-71 | TEXACO URSA SP-30 |
| | 38 | 1/6 | 83 | 5038 | WATER PUMP | ENGINE CAT D-315 | TEXACO URSA SP-30 |
| | 21 | 1/6 | 83 | 5021 | WATER PUMP | ENGINE DDA 3-71 | TEXACO URSA SP-30 |

**Figure 10.30** Confidence factor analysis. (Courtesy of Spectron International, Inc.)

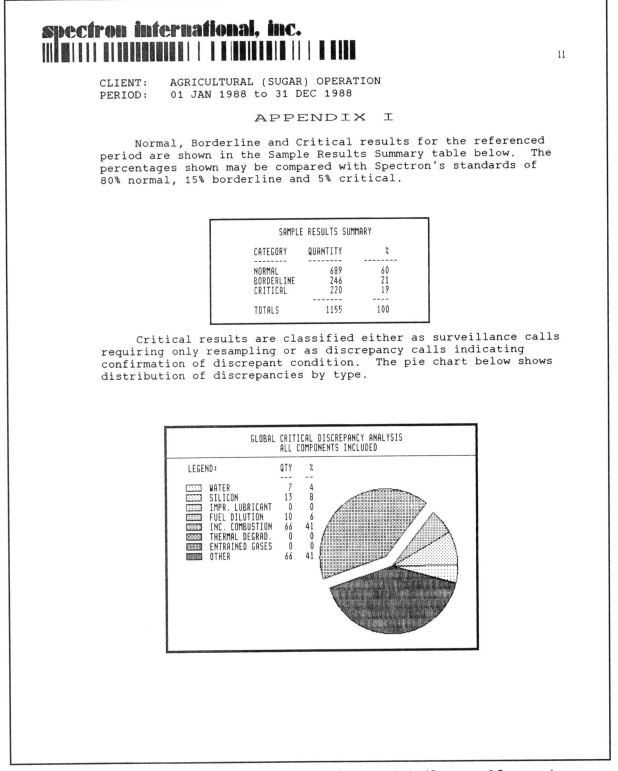

Figure 10.31  Confidence factor analysis. (Courtesy of Spectron International, Inc.)

16

```
CLIENT:    AGRICULTURAL (SUGAR) OPERATION
DIVISION:  WATER CONTROL
PERIOD:    01 JAN 1988 to 31 DEC 1988

               CONFIDENCE FACTOR ANALYSIS

        COMPONENTS WITH CONFIDENCE FACTOR OF 70% OR LESS
```

MON:  9  5009  WATER PUMP  ENGINE  DEUTZ AF4L912  TEXACO URSA SP-30  CFAC:  60

| DATE | CL | CD | DISCREP | FE | AG | AL | CA | CR | CU | MG | NA | NI | PB | SI | SN | TI | B | BA | P | MN | MO | V | ZN |
|------|----|----|---------|----|----|----|----|----|----|----|----|----|----|----|----|----|----|----|----|----|----|----|----|
| 23 Dec 1988 | W | 57 |         | 32  | 0 | 6   | 251  | 2   | 4   | 3  | 16 | 1 | 5    | 8   | 0 | 0 | 0 | 1 | 140 | 2 | 0 | 1 | 774 |
| 12 Oct 1988 | W | 57 |         | 41  | 0 | 4   | 249  | 0   | 3   | 3  | 19 | 0 | 5    | 9   | 0 | 0 | 0 | 0 | 122 | 0 | 0 | 0 | 998 |
| 29 Aug 1988 | R | 11 | SILICON | 141 | 0 | 29#305 | 7 | 12 | 10 | 20 | 2 | 28* | 25* | 0 | 0 | 2 | 2 | 169 | 3 | 4 | 1 | 998 |
| 13 Jun 1988 | W | 57 |         | 39  | 0 | 8   | 267  | 2   | 5   | 8  | 21 | 0 | 13   | 8   | 0 | 0 | 1 | 0 | 151 | 1 | 1 | 0 | 998 |
| 18 Apr 1988 | R | 11 | SILICON | 196*| 0 | 37#419 | 11* | 30* | 17 | 29 | 1 | 67# | 20 | 0 | 0 | 1 | 6 | 186 | 3 | 0 | 1 | 998 |

MON:  10  5010  WATER PUMP  ENGINE  DDA 3-71  TEXACO URSA SP-30  CFAC:  50

| DATE | CL | CD | DISCREP | FE | AG | AL | CA | CR | CU | MG | NA | NI | PB | SI | SN | TI | B | BA | P | MN | MO | V | ZN |
|------|----|----|---------|----|----|----|----|----|----|----|----|----|----|----|----|----|----|----|----|----|----|----|----|
| 15 Dec 1988 | R | 20 | INC CMBSTN | 84  | 0 | 1 | 212 | 6   | 7  | 3 | 17 | 2 | 4  | 9  | 7  | 0 | 2 | 0 | 126 | 2 | 1 | 2 | 714 |
| 12 Oct 1988 | W | 57 |            | 55  | 0 | 0 | 230 | 3   | 5  | 2 | 29 | 0 | 2  | 10 | 11 | 0 | 0 | 0 | 105 | 0 | 0 | 0 | 952 |
| 14 Sep 1988 | R | 21 | INC CMBSTN | 55  | 0 | 1 | 204 | 4   | 5  | 2 | 17 | 1 | 3  | 13 | 9  | 0 | 1 | 0 | 137 | 2 | 2 | 3 | 816 |
| 29 Jul 1988 | W | 57 |            | 31  | 0 | 3 | 240 | 2   | 8  | 9 | 17 | 0 | 2  | 6  | 0  | 0 | 2 | 0 | 137 | 2 | 1 | 0 | 920 |
| 9 May 1988  | W | 57 |            | 44  | 0 | 0 | 266 | 6   | 5  | 6 | 32 | 0 | 2  | 6  | 3  | 0 | 1 | 0 | 135 | 1 | 0 | 0 | 855 |
| 23 Mar 1988 | R | 20 | INC CMBSTN | 153*| 0 | 5 | 198 | 18* | 20 | 5 | 25 | 0 | 6  | 13 | 16 | 0 | 3 | 0 | 100 | 2 | 0 | 0 | 732 |

MON:  15  5015  WATER PUMP  ENGINE  DDA 4-71  TEXACO URSA SP-30  CFAC:  0

| DATE | CL | CD | DISCREP | FE | AG | AL | CA | CR | CU | MG | NA | NI | PB | SI | SN | TI | B | BA | P | MN | MO | V | ZN |
|------|----|----|---------|----|----|----|----|----|----|----|----|----|----|----|----|----|----|----|----|----|----|----|----|
| 17 Nov 1988 | R | 21 | INC CMBSTN | 270*| 0 | 3 | 205 | 8   | 7 | 3 | 18 | 0 | 7 | 10 | 14 | 0 | 1 | 0 | 113 | 3 | 0 | 0 | 773 |
| 23 Sep 1988 | R | 21 | INC CMBSTN | 213*| 0 | 0 | 213 | 9   | 6 | 3 | 18 | 2 | 8 | 9  | 9  | 0 | 2 | 0 | 147 | 3 | 1 | 2 | 783 |
| 9 May 1988  | R | 21 | INC CMBSTN | 217*| 0 | 3 | 220 | 27# | 5 | 8 | 16 | 0 | 4 | 12 | 9  | 0 | 0 | 0 | 124 | 3 | 0 | 0 | 664 |
| 7 Mar 1988  | R | 21 | INC CMBSTN | 372#| 0 | 0 | 203 | 41# | 5 | 3 | 15 | 1 | 7 | 12 | 7  | 0 | 1 | 0 | 120 | 5 | 0 | 3 | 729 |

```
Notes: 1) Asterisk adjacent to reading indicates threshold exceeded.  Pound symbol (#)
          indicates threshold exceeded by 100% or more.
       2) Calcium (CA) and Phosphorus (P) readings are reduced by a factor of 10.
```

**Figure 10.32** Confidence factor analysis. (Courtesy of Spectron International, Inc.)

duty system was needed. A major filter manufacturer was contacted and was able to provide a system tailored to their make and model of truck. The problem immediately cleared up. Oil and filter change intervals were extended to one month. This in itself was a savings, cutting the cost of oil and filters by 50 percent. A much greater savings was realized by eliminating the need for costly engine overhauls and the associated cost of downtime. Each new air intake system cost approximately $150 dollars and was paid for within one year by the savings in oil and filters. The trucks now have over three years of service and the engines are still in top condition.

Mechanical systems integrity management is another way of staying abreast of equipment maintenance. It should be obvious by now that this is a management tool used to control the maintenance function. It is designed not for the maintenance worker but for managers and supervisors. The system changes maintenance from a crisis reaction function to one of organized control of equipment through sound service and failure avoidance.

The technology is available and should be used to full advantage by those companies that wish to run their operations at peak efficiency and gain the most from their capital investments. It is not enough to check occasionally the quality of used lube oil. The integrity of the mechanical components must be monitored constantly. Managers who do not make use of the latest available technology will find themselves struggling to maintain control over their domain.

# chapter 11

# Applications of Tribology-Based Monitoring

In Chapter 10 we have shown how a machinery on-condition monitoring system can be integrated into an existing master plan. It thus becomes an extension of the system and an invaluable tool for management. The illustrations of forms and the various examples of reports have been taken from a typical agricultural enterprise. There are numerous other industries that make use of this type of system, and some are described later in this chapter.

The format of the forms and reports are essentially the same for all industries. There will be marked differences in the types of machinery and components monitored. These will be reflected in the monitor control lists. Unit condition reports, batch summary reports, monthly evaluation reports, and special reports use the same format for all industries. Special studies and reports, in addition to those already described, are tailored to the unique requirements of each industry.

Before going further, we should discuss how tribology is used in maintenance management and monitoring techniques. Tribology is the science of friction, wear, and lubrication. It has played a major role in the design of modern machinery and lubricants. To derive the full benefit from this technology, it should also be applied to maintenance, for it is this activity that will ultimately determine productivity and life of capital equipment.

Over the past decade, there have been developed many new techniques dealing with condition monitoring, yet the field of maintenance management has not reached its potential as a modern science. The fault is not with the techniques but with the general philosophy concerning maintenance.

Condition monitoring has evolved through three main categories: debris monitoring, wear monitoring, and tribology-based monitoring. *Debris monitoring* involves the analysis of gross metal particles, generally considered to be visible particles of 40 microns and larger. Techniques and devices applied to this type of monitoring include magnetic sump plugs and chip detectors, acid tests for the identification of metallic particles found during filter and strainer inspections, spectroscopy, and ferrography.

The presence of gross metal indicates that a component, or components, have already suffered major distress and/or failure. To further determine the composition of the particles by various testing techniques is of little use to the maintenance manager. These findings are all *resultant,* not *causal,* factors of mechanical damage. Further damage may be averted by shutdown and repair, but the time for preventive maintenance has long since passed.

The argument that debris monitoring is useful for revealing the failure mode, such as galling, scuffing, or spalling, is of little value except to academics. This is analogous to treating the symptom and not the disease.

However, many still advocate this technique's application to on-condition monitoring. The problem here is that it is nearly impossible to secure representative samples that contain representative concentrations of metallic particles. Without this, there is no basis for comparisons over an extended time. Without qualification, the rate of particle generation is lost and, consequently, so is any opportunity to measure the immediacy of failure.

*Wear monitoring* is generally defined as the analysis of wear particles suspended in the used lubricant. Particle size varies with the viscosity of the lubricant in which it is suspended. The upper limit is usually considered to be 10 microns. Spectroscopy is the analytical technique commonly used and includes atomic emission, atomic absorption, x-ray fluorescence, and inductively coupled plasma. Each has its advantages and disadvantages. Wear metals, lubricant additives, and certain contaminates are detectable. Because the elements are quantifiable, concentration trends can be established by continuous monitoring. These trends can indicate if a continuing wear condition exists, its wear rate, and the urgency of the problem. This technique provides more information faster and sooner than debris monitoring, but it still is observing particles. So it is still part of the philosophy of failure avoidance and reactive maintenance.

Oil analysis programs are the most common application of wear monitoring. Oil analysis, as applied here, is a misnomer because the oil is treated primarily as a vehicle carrying the wear particles. Some American Society for Testing and Materials (ASTM)-based tests are performed but they are more physical than chemical. They typically include viscosity, total base number (TBN), total acid number (TAN), total solids (by blotter or centrifuge), water (crackle test or KF titration), and flash point.

These tests were originally devoloped for use with straight mineral oils and not for modern, additive-enhanced lubricants. Tests have been modified for use with modern lubricants, with the result that they no longer measure the conditions for which they were originally designed. In addition, they are no longer appropriate for modern sophisticated lubricants.

Wear monitoring, by itself, thus has provided little new or pertinent information to the maintenance manager. True, it does provide more information much earlier and indicates trends that help the maintenance department make adjustments to alleviate problems. However, like debris monitoring, it forces management to react to the effects of mechanical stress without knowing the basic cause of the problem.

*Tribology-based condition monitoring* does provide an alternative to "reactive" maintenance. Tribology offers a perspective of the cause and effect of mechanical integrity. It enables maintenance to be focused on the control and management of the causes of friction and wear rather than the results of these two factors. This maintenance approach combines the most modern analytical techniques for the detection of tribologically significant factors with the knowledge of tribology to interpret them. This, coupled with mechanical experience, can translate these findings into specific maintenance actions and programs.

Mechanical systems integrity management is just such a program. It is a computerized system of applied tribology and advanced maintenance management. The data source for this program is the used lubricant from each component being monitored. The oil sample contains not only the metallic wear particles, but more important, virtually all of the tribologically significant factors which determine both the lubricant and mechanical integrity of the system.

The flowchart, Fig. 11.1, shows how information is passed from each workstation to a central computer. It also identifies the latest sophisticated analytical equipment, as used by one consulting laboratory. Of this equipment, differential infrared is the most important because of its ability to detect essential tribology factors. These factors are, for the most part, undetectable through traditional physical tests designed for simple mineral oils.

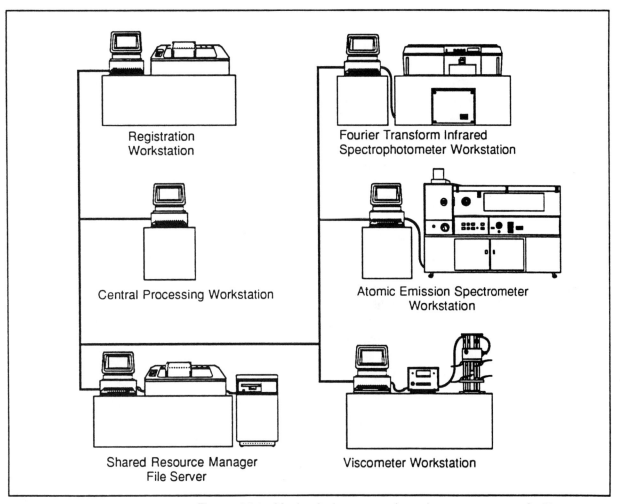

**Figure 11.1** Mechanical systems integrity management facility. (Courtesy of Spectron International, Inc.)

A comparison of 2000 data points is made between the spectra of the used oil and the unused reference oil. Differences indicate molecular (chemical) changes in the used oil due to contamination and/or degradation. These reflect actual conditions of both lubricant and mechanical integrity.

Sample measurements from each workstation are automatically transferred to a central computer, which then automatically calls up each component's file, updates it with the current results, then evaluates and interprets the results to produce a maintenance recommendation. The logic flow of the computer's interpretive program is based on years of experience in maintenance management, monitoring, and applied tribology being preprogramed into it.

Human interpretation is used to back up the computer's interpretation, especially in gray areas. When the client and consulting laboratory are working closely together trying to solve problems or to observe the results of initiated changes, human supervision of results must be used. Where there is a degree of change that might swing a unit condition report from being classified borderline to critical, the consultant might wish to override the computer and leave the unit sample classification as borderline. This may occur where a few more short-cycle samples might be needed to verify if a change in the maintenance procedure is beginning to show positive results.

Despite all this technological development and progress, a recurring question is asked: Is condition monitoring cost-effective? If it is accepted that the major responsibility of management is to promote the most cost-effective operation possible, it follows that maintenance management must meet the same standard.

During the last 30 years, experience in this field has shown that proper selection and application of condition monitoring techniques to programs of mechanical systems integrity management have been extremely cost-effective in general and in advanced programs have reduced disassembly inspections and repairs to less than 1 percent of all components monitored. Infrared spectrophotometry analyzes the qualitative difference between used and unused oil, thus allowing the use of lubricants until they are no longer in like-new condition. The average extension of useful oil life is generally sufficient to make a complete monitoring program cost-effective.

Some applications of this type of advanced on-condition monitoring have been discussed in previous chapters. The following list illustrates the diversity of industries that have benefited from this technology: mining, mechanized agriculture, construction and contracting, maritime (motor vessels, tankers, tugs, etc.) municipal transport (buses, trucks, etc.), long-haul trucking, ports authorities, taxi companies, power plants (municipal and private), railroads (diesel locomotives), alumina refining, cement plants, oil refineries, chemical plants, pharmaceutical production plants, aircraft, U.S. military. For a further insight into the types of components these industries monitor, some examples are briefly described below.

## *ALUMINA PLANT*

The monitor control list is broken down into sections. This is done so that each division can be compared with others in the various reports. Supervisors of each division have their own feedback and need not be concerned with reports dealing with the other divisions. Management is the most interested in the divisional comparisons and ratings.

In this example, under "Plant Equipment" are listed gear reducers, bearings, hydraulic couplings, and hydraulic drives. These are components of feeders, conveyors, ball mills, agitators, pumps, holding tanks, and mud settlers. "Power House Equipment" includes engines, turbine and compressor crankcases, turbine gears, gearboxes, pumps, bearings, and reservoirs. "Mobile Equipment" includes engines, transmissions, hydraulic systems, gears, and final drives. These components are from traxcavators, wheel loaders, forklift units, dempsters, cranes, welding machines, dozers, and emergency fire pumps. "Coal Handling" monitors gearboxes and housings used on crushers, reducers and conveyors. Finally, the "Coal Boiler" section monitors turbine pumps, fan bearings, air preheater gearboxes and bearings, bowl mill bearings, air compressor crankcases, and screw air compressor reservoirs.

## OIL REFINERY

A typical monitor list would include units and their components listed by divisions, with short titles or letter descriptions, also shown. This might include the following notations: utilities, MEK, maintenance, hydrogen, LHT, crude, WWTA, dock, TK farm, GOD, gas reform, and cool tower. Storage tanks would also be listed under "new oil storage." Obviously, space prevents the listing of all monitored components for such a large installation.

## MOTOR VESSELS

Components monitored include everything from the main propulsion system through feed pumps in the engine rooms. Other equipment would be found in the refrigeration and air conditioning systems, compressed air system, steering system, and emergency diesel generators. Storage tanks, lifeboat engines, winches, drags, hoists, and torgue converters are among the many monitored components. Supertankers have most of these same monitored units, plus cargo pumps, autopilot systems, and others unique to their operation and service.

## MINING INDUSTRY

The most important units monitored in open-pit mines are the Haulpaks, which represent a huge capital investment. These units have engines, transmissions, differentials, and hydraulic systems that must be strictly monitored. Construction and earth-moving equipment are also monitored along with all storage facilities for fuel and lubricating oils, be they stationary or mobile units.

## PHARMACEUTICAL PLANTS

The production and auxiliary machinery for the pharmaceutical industry must be closely monitored to prevent any work slowdown or stoppage. If, for some reason, a unit should shut down during a production run, the whole batch could be ruined and lost. Unit components are monitored by sections: for example, fermentation, finishing, extraction, LSS plant, cogeneration oprations, cogeneration maintenance, process air system, compressed air system, cooling tower system, chilled water system, and glycol system. In addition, new oil storage tanks are checked continuously.

These examples of monitored components only scratch the surface but do serve to illustrate the diversity of both machinery and industry that use on-condition monitoring and mechanical systems integrity management.

A few case histories will serve to illustrate the value of this advanced technology when applied to preventive maintenance. As a result of the program, each case shows how the source of a problem was identified and what steps were taken either to remove the cause or to adjust the maintenance procedure to effectively counteract the basic cause. In some cases, the action taken had to be tempered by the practicalities or economics of a particular situation.

## CASE 1: A MAJOR METROPOLITAN TRANSIT AUTHORITY

As with most such organizations, this one is heavily subsidized, and financial limitations seriously affected operations and maintenance. A program was put in place to gain a measure of cost-effective control. Their rapidly deteriorating fleet of over-aged coaches had 65 percent of the units classified as critical during the first months of the program.

Four years later, only 5 percent of the 400 buses would fall within this category in any given month. This enabled management to extend the lube service interval 100 percent, thus reducing maintenance expenditures on labor and materials.

These results were achieved mainly by gathering the appropriate information concerning the fleet which enabled management to operate more efficiently. From the start, analytical data revealed major inadequacies in the lubricant used. A different engine oil with a more highly constituted additive package was substituted with excellent results. However, continued monitoring revealed a chronic problem of incomplete combustion throughout the fleet.

It was known that low compression and worn fuel injection systems were the cause. There simply were not sufficient funds available to perform all the required overhauls. Consequently, the only option available was to change from diesel fuel to a lighter, more volatile distillate such as kerosene. The resultant improvement in combustion reduced exhaust emissions to an acceptable level. The slight reduction in calorific value did not adversely affect engine performance. The success achieved with this fuel was so positive that it continues to be used.

The long-term effectiveness of this program is evidenced by the fact that, today, the extended ratings for this fleet are predominately in the good-to-excellent category.

## CASE 2: OPEN-PIT MINING

To determine the feasibility of extending oil change intervals from 125 hours, management requested an evaluation program. The data acquired through tribology-based condition monitoring plus analysis of diesel fuel, over a period of one year, formed the basis for the evaluation and recommendation.

As long as the fuel sulfur stayed at 0.5 percent, the Caterpillar crawler and loader units could be expected to realize safe service intervals of 375 to 500 hours. This was an increase of between 300 and 400 percent. The 60-ton Haulpaks could have their oil-change intervals extended to between 1500 and 3000 hours, which represented an increase of between 1200 and 2400 percent. The net savings in oil, filters, and service labor amounted to an approximate annual savings of $100,000. This was predicated on the continued, tightly controled, routine service maintenance program in force at the time.

High-sulfur-content fuel has been a problem throughout the world. One approach to a solution has been to revert to an overbased lubricant with a TBN 20 times the percent of sulfur contained in the fuel and reduce the oil change interval to as little as 50 hours. This is an extremely expensive measure and totally inappropriate in cases where two-cycle engines are involved.

One mine solved this problem very easily and economically through tribology-based condition monitoring and controlled testing of alternative methods. The result was to stay with their standard lubricant, with a TBN of 7, but drop from a SAE 40 to a SAE 30 grade. The fuel remained unchanged. The oil consumption was thus intentionally increased. The constant addition of makeup oil provided a replenishment of additives, including alkalinity, to maintain the oil at peak effectiveness. Weekly monitoring of this approach showed no degradation due to the fuel. In fact, it was discovered that the service intervals could actually be extended due to the excellent condition of the crankcase oil charge. This served to offset the added cost incurred by the higher rate of lube oil consumption.

One mine suddenly began to suffer accelerated piston ring and liner wear in the engines. Analytical results from the program also noted a change in the additive blend of the lubricant. It was observed that the antiwear additive concentrations were greatly reduced from those in the lubricants supplied previously.

The lubricant supplier was consulted and another lubricant with a higher antiwear additive content was chosen. Condition monitoring showed an immediate improvement in the wear condition of the engines. A year later, statistical wear profiles were made of these engines to compare the effects of the two lubricants. Not only had the piston ring and liner problem been resolved, but the overall wear profiles of the engines had been reduced by about 53 percent.

When an additive package is well matched to the task, the results are superior lubrication. Conversely, a mismatch can cause accelerated wear. Tribology-based condition monitoring enables a consulting laboratory to observe this phenomenon.

## CASE 3: OIL REFINERY

Consulting labs have for years observed the reactions of various types of mechanical systems and operations to the changing, or mixing, of lubricants. Perhaps the most notable case was that of an oil refinery, where one brand of lubricant was changed for another under a new purchasing agreement.

Existing stocks of the original oils remained on hand at the time the agreement went into effect. The decision was made to use these stocks only in certain systems until they were depleted. The remaining mechanical systems were drained and the new oil added. A concerted effort was made not to mix lubricants, which obviously would have courted disaster.

Despite these efforts, only those systems that had not been switched to the new lubricants maintained their prior stable patterns of mechanical and lubricant integrity. The others required increased surveillance and maintenance attention such as might be expected with a new plant startup. All systems within the refinery went through the same procedure as the old lubricant stocks became depleted. The total period of instability caused by this change lasted approximately 18 months, with accompanying increased maintenance costs.

Modern lubricants are much better than they were 20 years ago but are also much more complex. Great care and sophistication are required for their selection and application.

## CASE 4: PERSONNEL AND MISCELLANEOUS FACTORS

Not all problems and solutions have to do with the technical side of fuels, lubricants, and their applications to machinery productivity. Continuous surveillance will often reveal flaws in personnel and/or management control and even machine design or fabrication. A few examples will serve to illustrate these factors.

Monitoring data and cost accounting data at a mine indicated serious problems with the Caterpillar equipment. Monitoring data disclosed the chronic presence of silicon in all components. A joint investigation revealed that this equipment was being serviced during the night shifts by production personnel rather than maintenance personnel. Through an oversight, the night crews had not been provided with lubrovans. As a consequence, the equipment was being serviced from oil drums and filler cans that were strewn around the floor of the pit. Within 30 days after this staffing problem was corrected, the reduction in silicon and abrasive wear was obvious. Within six months, cost accounting data were back to normal.

A case of overmaintenance was identified as the cause of abrasive wear problems in the final drives of crawler tractors. It was the practice to check oil levels every day. Each time, dirt was introduced through the plug hole. Since any loss of lubricant could occur only through the outboard seals, it was recommended that the daily service inspection of these components be limited to a visual check of the outboard seal area for signs of leakage. The oil level would be checked once a month. After this procedure was adopted, the rate of abrasive wear dropped.

In another case, the consulting lab detected traces of gasoline in the crankcase oil samples taken from the diesel engines of a client's tractor fleet. Needless to say, this caused some eyes to pop and heads to shake. The cause was finally attributed to the client's sampling technician. He had been wiping off his sampling equipment, notably the pickup tube, with a rag that had traces of gasoline in it. Periodically he would clean his hands with gasoline and dry them with the rag. As soon as he began using clean paper towels, the phantom problem disappeared.

The last example is a case of a defective major component. A nearly new diesel-powered wheel tractor showed chronic silicon and resulting wear. Nothing the client's maintenance people did seemed to help. The lab at first suspected that personnel were not doing a thorough job of cleaning, inspecting, and checking the air intake system. The unit was put on 50-hour surveillance monitoring and still nothing could be done to change the situation.

In desperation, the maintenance manager dismantled and reassembled the intake system and installed a new air filter and separator. He then flushed the crankcase with new oil and added a new charge of oil and a new oil filter. Next, he drove the tractor into an empty section of a warehouse and ran it there for 50 hours, then took an oil sample. The results were unchanged! By this time, client and lab consultants were completely baffled, so they decided to tear down the engine. What they discovered was a defective block casting. Casting sand was seeping from pores in the metal into the oil. Either the special paint on the inside surfaces of the block, designed to prevent any residual sand seepage, was insufficient or the casting was too porous. When presented with this evidence, combined with their own physical examination, the manufacturer supplied a new engine under warranty.

In industries employing large quantities of diverse equipment, it is difficult to see beyond the immediate day-to-day problems associated with individual units and attain a broader, long-term, management perspective. Mechanical systems integrity management, based on tribology, can provide that prospective. If used with imagination, the benefits can be enormous.

These cases help to illustrate the wide diversity of the system throughout industry. In every case, direct savings from using the system are greater than the cost of operating it. However, the biggest advantages occur when the following stages are finally achieved:

1. Strict control of the maintenance operation
2. Strict control of supervision
3. Management concentration on the small number of chronic problem areas
4. Management evaluation of the total operation
5. Assurance that fuel and lubricants are to specification through preuse monitoring
6. Reduced downtime, spares inventory, maintenance costs, and increased service life of the capital equipment

# chapter 12

# Summary

- Philosophy of Maintenance
- Evolution of Modern Maintenance
- Overview

This book has set forth the basic steps needed to attain a high level of maintenance management. It has also stressed the goal every company should aim for: the greatest productivity from the capital equipment at the lowest cost.

Maintenance can no longer be considered as an isolated technical activity. It must be an important part of management. Because of the cost and importance of the capital equipment, the maintenance organization must be granted a strong voice in management. It can only do this if it is prepared to function technically and effectively by using a sound management approach to maintenance of machinery and equipment. The maintenance function must not be thought of as one of repair with some occasional and informal service. This approach is simply a reaction to breakdown and damage.

Originally, maintenance was perceived as merely failure repair. A good maintenance organization was one that could make repairs quickly and get the equipment back into operation. This held true until World War II, when reliability became a major factor. Steps were taken to avoid failures of equipment through the use of scheduled maintenance. Small failures were corrected in hopes of possibly avoiding catastrophic failures somewhere down the line. The philosophy of maintenance has since passed through a number of stages. They can be labeled as failure avoidance, failure prevention, and finally, on-condition monitoring. The latter brings maintenance up to the level of today's technology and philosophy of progressive management.

To help the maintenance organization cope with the complexities and diversities of modern equipment requires that those in charge apply the most up-to-date management skills and seek out the latest aids available for equipment control. It is virtually impossible for most companies to employ enough technical people to have expert coverage of all the different types of machinery and their components. To attempt to do so is not only costly but based on an incorrect concept of modern maintenance. That outdated concept holds that maintenance has as its first priority the repair of equipment.

Maintenance should be considered as the husbandry, or judicious care, of equipment. To be economically feasible, capital resourses must be employed at full capacity for as long as possible, meaning for the full life of the equipment. There is no excuse for allowing failures caused by ignorance, neglect, or stupidity. No one can exercise control over accidents or acts of God, but it is possible to control the condition and operation of equipment. The key word is *control* and without it, all is lost.

To establish control, the basic steps required have been described in the preceding chapters, starting with organization. A streamlined organization, staffed by technically capable personnel, is a basic necessity. Once an effective organizational structure is established, the next step is the selection of the right people. It is vitally important that the best people be selected for key positions. They should be innovative and enthusiastic in addition to being technically qualified.

The next step is to establish a basic master plan covering equipment maintenance and service. This is nothing more than identifying the equipment that is on hand and the service requirements for each unit. To accomplish this task, a series of steps have been set down to aid in simplifying the process. As each step is completed, more order will be established in what may have seemed, at the start, to be a hopelessly confusing situation. Once order is established, it is easy to exercise control. This master plan is the foundation on which everything in the maintenance department is based.

Types of companies, their equipment, and operations may vary but the basic approach to modern machinery maintenance remains the same. When these fundamentals are mastered, the data collected can be put to additional use to enhance the efficiency and economy of the operation. Computers can be used to aid both staff and management by easing the work load and presenting data in forms that help simplify the budget and planning processes. Computers are useful when projecting trends and equipment requirements on both the department and company levels.

The basic master plan, as described through Chapter 8, will allow the maintenance manager and staff to have full control over the service and repair functions. It will also enable the manager to collect accurate cost data that will form the basis for future studies and management decisions.

Chapters 9 through 11 carry maintenace to the next logical level, in step with the latest available technology. The philosophical acceptance of maintenance as being fundamentally remedial in nature is no longer feasible. Maintenance must be considered to be primarily a tightly controlled approach to the conservation of machinery through preventive techniques. When this is put into practice, the managers and staff can concentrate on long-term planning. They can take time to study problem areas and formulate lasting solutions rather than spending the bulk of their efforts in crisis management. The effectiveness of management in maximizing productivity and conserving capital assets is a key factor in attaining business success.

Mechanical systems integrity management goes a long way toward achieving this goal. The system also enables top-level management to become familiar with the maintenance of their capital equipment and helps bring the maintenance organization into full partnership with the other company departments.

It is sometimes thought that this type of program is too sophisticated for use in operations where the level of skill and preparation of the labor force are low. Quite the opposite is true. These programs are designed for use by supervisory and managerial personnel and are not influenced by the technical level of wage earners. In fact, the lower the level of competency, the greater the need for continuous feedback of information and control.

This brings us back to that most important factor, control. To exercise effective control, an organization must be thoroughly familiar with its operation. In this book we have attempted to present a simplified way of achieving this by explaining, in detail, the necessary steps for setting up an effective service and maintenance organization.

# Index

# List of Illustrations

**CHAPTER 4**

**CHAPTER 5**

**CHAPTER 6**

**CHAPTER 7**

**CHAPTER 10**

**CHAPTER 11**